The cycling of elements such as carbon and nitrogen is of central importance in ecology, particularly when humans are causing changes to element cycles on a global scale. In this book a rigorous mathematical framework is developed to analyse how element cycles operate and interact in plants and soils, forming the foundations of a new ecosystem theory. From a few basic equations, powerful predictions can be generated covering a wide range of ecological phenomena related to element cycling. These predictions are tested extensively against field and laboratory studies of agricultural and forest ecosystems.

This groundbreaking work will be of interest to graduate students and researchers in theoretical ecology, biogeochemistry, forestry, agronomy, soil science and plant ecology.

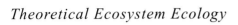

Theoretical Ecosystem Ecology

Theoretical Ecosystem Ecology

Understanding Element Cycles

Göran Ågren & Ernesto Bosatta

CAMBRIDGE
UNIVERSITY PRESS

PUBLISHED BY THE PRESS SYNDICATE OF THE UNIVERSITY OF CAMBRIDGE
The Pitt Building, Trumpington Street, Cambridge CB2 1RP, United Kingdom

CAMBRIDGE UNIVERSITY PRESS
The Edinburgh Building, Cambridge CB2 2RU, UK http://www.cup.cam.ac.uk
40 West 20th Street, New York, NY 10011–4211, USA http://www.cup.org
10 Stamford Road, Oakleigh, Melbourne 3166, Australia

First published 1996
Reprinted 1998

Printed in the United Kingdom at the University Press, Cambridge

Typeset in Times New Roman 10/13 pt

A catalogue record for this book is available from the British Library

Library of Congress Cataloguing in Publication data

Ågren, Göran I.
Theoretical ecosystem ecology/Göran Ågren & Ernesto Bosatta.
    p. cm.
Includes bibliographical references and index.
ISBN 0 521 58022 6 (hbk)
1. Biogeochemical cycles – Mathematical models.   2. Plant
ecophysiology – Mathematical models   3. Soil ecology – Mathematical
models   I. Bosatta, Ernesto.   II. Title.
QH344.A35 1996
574.5'222 – dc20 96–18908 CIP

ISBN 0 521 58022 6 hardback
ISBN 0 521 64651 0 paperback

To Elisabeth and Toty

Contents

Preface   xii

List of important symbols   xiv

PART I. PRELUDE   1

1   **Introduction**   3

1.1   Theoretical ...   3

1.1.1   The language   3

1.1.2   The tool   4

1.1.3   The model   6

1.2   ... Ecosystem ecology   6

1.2.1   The problems   7

1.3   Setting a perspective   9

2   **Element cycling**   12

2.1   Elemental distribution   12

2.2   Carbon and nitrogen   13

PART II. THE SOIL   17

3   **Theory for homogeneous substrates**   19

3.1   Main concepts   19

3.2   Carbon and nitrogen dynamics   20

3.3   A particular solution   26

3.4   Lag-time effects   29

3.5   From single litter cohorts to the soil   30

4   **Theory for heterogeneous substrates**   34

4.1   Substrate quality   35

4.2   Basic equations for carbon   35

4.3   Two particular solutions   39

4.4   The moment expansion   42

4.5   Basic equations for N, P and S   45

4.6   One litter cohort   47

4.7   Models for decomposer functions   51

4.8   Model I   52

4.8.1   One litter cohort   52

4.8.2	Several litter cohorts	54
4.9	Model II	55
4.9.1	One litter cohort	57
4.9.2	Several litter cohorts	58
4.10	Comparison with other approaches	59
5	**Carbon and nitrogen - applications**	**61**
5.1	Single litter cohorts	61
5.2	Steady state of soil organic matter	68
5.3	Correlation between carbon and nitrogen turnover	70
5.4	Steady state of a Scots pine forest	70
5.5	An agricultural application	72
5.6	Decomposer biomass	73
6	**Carbon, nitrogen, phosphorus and sulphur - applications**	**79**
6.1	C-N-P-S interactions	79
6.2	C, P and N dynamics in single litter cohorts	80
6.3	Stabilisation of C, N, P and S in the soil	82
6.4	C, N, P and S mineralisation	85
7	**Interactions with abiotic factors**	**91**
7.1	The problems	91
7.2	Mineralisation-immobilisation in single litter cohorts	92
7.3	N retention in the soil organic matter	97
7.4	Perturbation in carbon accessibility	100
7.5	Variable decomposer growth rate	107
PART III. THE PLANT		**111**
8	**Theory for plant growth**	**113**
8.1	Nitrogen productivity	114
8.2	Nitrogen productivity or photosynthesis and respiration	118
8.3	Different uptake models	120
8.3.1	General uptake rate	120
8.3.2	Exponential uptake	124
8.3.3	Fixed amount	126
8.3.4	General uptake rate - revisited	130
8.4	Nitrogen productivity and other nutrients	131
8.5	Variable P_N	133
8.6	Comparison with other approaches	137

9	**Plant growth - applications and extensions**	139
9.1	Empirical evidence for nutrient productivities	139
9.1.1	Exponential growth	139
9.1.2	Non-exponential growth	141
9.2	Nitrogen productivity and light	142
9.3	Root:shoot ratios	148
9.4	Nutrient use efficiency and the Piper-Steenbjerg effect	151
9.4.1	Nutrient use efficiency	151
9.4.2	The Piper-Steenbjerg effect	152
	PART IV. THE ECOSYSTEM	157
10	**Elements of an ecosystem theory**	159
10.1	The general terrestrial ecosystem equation	159
10.2	Ecosystem stoichiometry	161
10.3	Comparison with other approaches	164
10.3.1	CENTURY, GEM, G'DAY	165
10.3.2	FORET, LINKAGES	165
10.3.3	MEL	166
11	**Ecosystems - applications**	167
11.1	Ecosystems and global change	167
11.2	Nitrogen saturation	170
11.3	Short rotation forestry	174
11.4	Other applications	176
11.4.1	Global change - revisited	179
11.4.2	Forest nutrient budgets	180
11.4.3	Acid depositions to forest ecosystems	183
12	**Quality - the bridge between plant and soil**	186
12.1	Components of quality	186
12.2	Plants and quality	191
Epilogue		193
Appendices		195
A.1	Some properties of the delta function	195
A.2	Numerical solutions to (4.7) and to the moment expansion	196
References		201
Solutions to selected problems		222
Subject index		229

Preface

Theoretical ecology textbooks almost exclusively discuss population and community ecology questions. Their counterparts in ecosystem ecology bear titles with models or modelling as key-words. This seems to indicate that the level of abstraction or understanding in depth in ecosystem ecology is lower than in other areas of ecology. We have written this book to show that a theoretical ecosystem ecology also exists.

The present book has two main themes: plants and soils. Each of these themes is played as variations on one fundamental tune. Once the reader has understood these fundamental tunes, he or she should be able to develop his or her own variations. Some of the variations we are playing introduce fundamental new ideas whereas others serve to train the reader in his/her own ability to improve around these themes. In a final part we give some examples of how these themes can be intertwined. Our tenet is that once the *key concepts* are found, a few general equations are sufficient to describe large classes of phenomena, but for each phenomenon the specifics have to be included through a model. The fundamental ideas, theories, are abstract objects. It is in their interpretation through models that observables are defined. It is our experience that ecologists are unaccustomed to think in abstract terms. However, we know that by using abstractions over and over again, they eventually appear as ordinary objects that can be manipulated and changed as easily, or even more easily, than their concrete counterparts. The message we want to convey in this volume is two-fold. It is, of course, factual knowledge, but, as important, the method by which this factual knowledge is expressed. One fundamental tune can be played in several keys - the same theory should be able to answer more than one question. The more powerful the theory is, the more questions can it answer, albeit needing different models for its interpretation.

This book is written by two ex-physicists, as might be guessed from the examples used. The structure of books in physics has largely influenced us. Yet, we do not think that ecology is ready for the style of the typical theoretical physics book which could develop without reference to a single item of empirical evidence. Ecological theories are yet not sufficiently well established to be accepted without empirical back-up. There are, therefore, ample references to and testing against empirical information in this volume. However, we believe that the formalism in itself is important, and understanding how to manipulate and adapt it to different situations is invaluable. This should be done without the distraction of perpetually looking at specific examples. We have therefore organised this book in such a way that in each section we start by introducing the formalism for that particular area in one or two chapters to illustrate its major features. The reader will, at this point, have to believe that the formalism can reproduce empirical experience. In the following chapters, we demonstrate how the formalism can explain a series of empirical results.

There are also a number of problems inserted in the text. These problems serve to identify derivations of results that we do not want to include in the text. The problems may also concern applications not covered by the text. To these latter problems we suggest answers at the end of the book.

The text is aimed at graduate students and scientists in ecology, soil sciences, and plant ecophysiology or any other student or scientist dealing with plants or soils and who wants a theoretical perspective on their field of research. The language of the book is mathematics. This is unavoidable as this is the best tool of the trade for logical reasoning in natural sciences and we have not seen scientists otherwise avoiding the most powerful tools available.

A large number of people have assisted in the development of this book. Several versions have served as lecture notes for courses that we have taught and we thank these students for their contributions in improving the text. A special thank you goes to Fredrik Wikström who proofread and commented on the whole text. Berit Lundén helped reformat a previous version. Bengt Olsson drew Figure 1.1 for us. Åsa Ågren helped redraw the figures. Finally, we thank all our colleagues who have through the years helped us develop an understanding of ecosystems.

The following publishers have granted their kind permission to reproduce material copyrighted by them: American Society of Agriculture, Annals of Botany, Blackwell Science Ltd., The Ecological Society of America, Elsevier Science Ltd, Elsevier Science - NL, John Wiley & Sons, NRC Research Press, Oikos, Physiologia Plantarum, PUDOC, The University of Chicago Press.

List of important symbols

A list of the most important symbols used along with the equation where they first appear is given. Some symbols are used with several different meanings. A number of additional symbols are used locally. Several of the symbols will also appear with indices, where in particular the following are used: 0 = initial value; C = carbon; L = litter type; N = nitrogen; p = perturbation; P = phosphorus; s = soil; S = sulphur; ss = steady state; v = vegetation.

Symbol.	Meaning	Typical dimension	Equation
a	Maximum nitrogen productivity	kg dw (kg N)$^{-1}$yr^{-1}	9.5
a	Age of litter cohort	d	3.30
a'	$P_N c_{N,min}$	d^{-1}	8.17a
b	Decrease in nitrogen productivity	ha (kg N)$^{-1}$yr^{-1}	9.5
B	Microbial biomass	g dw m^{-2}	3.6b
c_N	Nitrogen concentration	g N (g dw)$^{-1}$	8.3
$c_{N,min}$	Minimum nitrogen concentration	g N (g dw)$^{-1}$	8.1
$c_{N,opt}$	Optimum nitrogen concentration	g N (g dw)$^{-1}$	Fig. 8.1
C	Soil carbon	g C m^{-2}	3.2
D	Dispersion function	—	4.2
e	Microbial efficiency	—	—
e_0	Parameter in microbial efficiency	—	4.41
e_1	Parameter in microbial efficiency	—	4.41
$\mathbf{E}$	Ecosystem operator	—	1.1, 10.2
f_C	Microbial carbon concentration	—	3.1
f_L	Leaf fraction	—	8.5
f_N	Microbial nitrogen concentration	—	3.2
F	Effect of inorganic N on immobilisation rate	—	7.8
g	General uptake function	—	8.16
g	Fraction of carbon remaining	—	3.16
h_c	Maximum of h_N	g N (g C)$^{-1}$	3.19
h_N	Amount of nitrogen relative to intial amount of carbon	g N (g C)$^{-1}$	3.17

Symbol	Description	Units	Reference
I	Input rate of soil carbon	$g\,C\,m^{-2}d^{-1}$	3.30
k	Specific decomposition rate	$d^{-1},\ yr^{-1}$	3.12a
m	Net mineralisation rate	$g\,N\,m^{-2}d^{-1}$	3.2
M	Microbial mortality rate	$g\,m^{-2}$	3.6a
n	Soil nutrient amount	$g\,m^{-2}$	4.31
N	Plant nitrogen	$g\,N\,plant^{-1},\ kg\,N\,ha^{-1}$	8.1
N	Soil nitrogen	$g\,N\,m^{-2}$	3.2
N_i	Soil inorganic nitrogen	$g\,N\,m^{-2}$	7.8
N_r	Reduced amount of nitrogen	—	8.20
P	Production rate of microbial biomass	$g\,m^{-2}d^{-1}$	3.1
P_N	Nitrogen productivity	$g\,dw\,(g\,N)^{-1}d^{-1},$ $kg\,dw\,(kg\,N)^{-1}yr^{-1}$	8.1
q,q'	Quality	—	4.1
$\hat{q}$	Average quality	—	4.21
Q	Photon flux density	$mol\,m^{-2}s^{-1}$	8.5
r	Nitrogen:carbon ratio	$g\,N\,(g\,C)^{-1}$	3.3
r_c	Critical nitrogen:carbon ratio	$g\,N\,(g\,C)^{-1}$	3.3
r_m	Maximal relative growth rate	d^{-1}	Fig. 8.1
R	Respiration rate	$g\,m^{-2}d^{-1}$	3.1
R_N	Relative uptake rate of nitrogen	d^{-1}	8.4
R_W	Relative growth rate	d^{-1}	8.3
s	Specific rate of abiotic immobilisation	d^{-1}	7.8
t	Time	$d,\ yr$	
t_c	Critical time	$d,\ yr$	3.18
u	Microbial growth rate	$g\,dw(g\,C)^{-1}d^{-1}$	3.11
u_0	Parameter in microbial growth rate	$g\,dw(g\,C)^{-1}d^{-1}$	4.42
U	Level of nitrogen uptake	—	8.16
W	Plant biomass	$g\,dw,\ kg\,dw\,ha^{-1}$	8.1
W_r	Reduced plant biomass	—	8.21
β	Parameter in microbial growth rate	—	4.42
η_n	Moment of dispersion function	—	4.14, 4.24
η_{10}	Parameter in first moment	—	4.43
η_{11}	Parameter in first moment	—	4.43
μ_b	Microbial mortality rate	d^{-1}	4.1
μ_n	Moment of carbon distribution	—	4.23
μ	Mortality of biomass	yr^{-1}	10.3, 10.4
ρ_C	Carbon distribution	$g\,C\,m^{-2}$	4.1

ρ_b	Carbon distribution in microbial biomass	$g\,C\,m^{-2}$	4.1
ρ_n	Nutrient distribution	$g\,m^{-2}$	4.30
$\vec{\rho}$	Ecosystem state vector	—	1.1, 10.1
σ^2	Variance of carbon distribution	—	4.22
τ	Time	d, yr	—

PART I

PRELUDE

Physics is mathematics
not because we know
too much about the physical world,
but because we know too little;
the only thing we can discover
are the mathematical properties
of the physical system.

(Bertrand Russell)

1

Introduction

This book is a theoretical treatise on element cycling in ecosystems.[1] It is centred around one concept and revolves around one equation. In the Introduction we discuss the meaning of key words related to its title, *Theoretical Ecosystem Ecology*. These words are: *language, tool, model* and *problem*.[2]

1.1   Theoretical...

A *theory* is a set of concepts (the *language*) linked by mathematics (the *tool*) and used to analyse specific *problems* by being translated through *models*.

1.1.1   *The language*

According to the above definition, the basic building blocks of a theory are the concepts. Mathematics is only the tool for working with these building blocks and no theory is stronger than the concepts on which it is based. It is generally believed that an increase in theory generality, i.e. an increase in its ability to explain an increasing number of problems, is related to the sophistication of its mathematics. But this is wrong. Our tenet is that theory evolution has to do with the "invention" of germinal concepts.

As an example of this let us consider Classical Mechanics. There is a clear increase in generality along the path from Galilei (early 17th century) to Newton (late 17th century) and to Hamilton (19th century). Galileian mechanics could solve the kinetics of falling bodies. Newton's equations allow us to solve the problem of motion in a central force field, of which the falling body is a special case. The Hamiltonian point of view allows us to solve a completely new series of problems, for example, the problem of attraction by two stationary centres. Hamiltonian mechanics is also of importance for its links with optics and quantum mechanics (20th century).

But the evolution Galilei→Newton→Hamilton is not related to mathematical sophistication but to the *conceptual evolution* acceleration→force→energy (acceleration was actually "invented" by Galilei). The step to relativistic mechanics was also a conceptual revolution, but more subtle: it implied a change in the symmetry of the universe. In Classical Mechanics the laws of physics are invariant under the group of Galileian transformations. By making the velocity of light a constant, Einstein made the laws of physics invariant under the more general group of Lorentz transformations.

One particular kind of concept concerns the so-called non-observables. An example comes from quantum mechanics, where the basic entity, the wave function, exhibits a number of characteristics making it unobservable. However, as long as these non-observables can be projected onto observables they may continue to play a fundamental part in the theory. Quantum theory, based upon an unobservable, has been extremely successful in explaining physical phenomena.

Thus, it is the increase in conceptual sophistication that leads to increased abstraction and generality in a theory. The increase in mathematical sophistication is a consequence of the increase in conceptual sophistication and not vice versa. The critical question is then how to "invent" the germinal concepts. Our tenet is that this is the point where the road of science intersects the road of art. Science does not evolve in the form of a soft, mechanistic path but rather by means of jumps of random length. It is just as impossible to write a manual about inventing germinal concepts as it is to write a manual about teaching Nijinsky how to dance.

1.1.2   *The tool*

In natural sciences the tool is mathematics.[3] The importance of mathematics is that it provides the theory with a dynamics, an explorative dynamics. Other points that make mathematics useful are:

Help in theory construction. Advanced theories are necessarily expressed in a formal language. Mathematics is today the best developed formal language. It provides us with a number of symbols, e.g. $+$, $=$, d/dt, etc., with which we can formalise our relations.

Accuracy. The use of mathematics necessitates more precise statements. Vague propositions like "the growth rate of a plant depends upon its leaf biomass," have to be replaced by more precise ones like "the growth rate of the plant is proportional to its nitrogen content." The accuracy implied is not a numerical one, i.e. the power to predict plant size to six digits, but more a qualitative one. The ambiguity of statements in ordinary language is reduced and even vanishes.

Deductive power. Science strives to summarise knowledge in powerful statements entailing as many consequences as possible. In ordinary language it is difficult, not to say impossible, to extract any elaborate consequences of a statement. This can only be done with a formal language.

Metatheoretical advantages. Demonstrations of inconsistencies and dependencies within a theory are best done in mathematical terms, cf. deduction.

Comparisons between theories. The accuracy of statements made in mathematical form avoids unnecessary conflicts over the interpretation of the statement. Unambiguous comparisons between the content of different theories are thereby facilitated.

Generality. The mathematical language is used in several scientific disciplines. Thus, the same mathematical analysis can often be performed for entirely different systems. By reinterpreting the symbols, earlier results can be used, thereby saving us much time and labour in finding solutions to a mathematical problem.

Simplification. Only relatively simple formulations permit us to deduce any results. This should be seen as an advantage, because it forces us to try to extract the essentials from a problem. We must be aware that already seemingly very simple mathematical formulations may entail extremely complicated consequences.[4]

One should be careful in this context not to confuse mathematics with computers (cf. **Simplification** above). Suppose today's computers had been available in the days of Kepler. Would it not have been much easier for the astronomers of those days to make complicated computer programs to calculate the epicycles found by Appolonius, predicting, certainly with high accuracy, the movements of the planets. Tycho Brahe was astonished by the accuracy with which predictions could be made within the Ptolemaic astronomy. Would it then have been possible to shift to a heliocentric world using the simple laws of Kepler if these had been found less accurate, as is likely to be the case with, e.g. Mercury, the movement of which could not be properly explained until the beginning of this century, when Einstein had developed relativistic mechanics? This should remind us that the goal of science is not prediction but understanding, and understanding is genuinely qualitative whereas prediction is quantitative. This also indicates that difficulties with the tool may stimulate the appearance of a conceptual revolution.

1.1.3   *The model*

In well-developed theories, language and tool evolve towards a very economic state in which the germinal concepts are combined by mathematics into a single equation and the constituent parts of the equation are symbols representing very general classes of relationships. If we want to study a particular problem, we select a particular element of the class of relationships (e.g. a particular function) and find the solution of the equation for this particular problem. We say that we have *translated* the theory *through* a model or that the model is a particular realisation of the theory.

The central equation of the theory can then be seen as an infinite collection of models where each projection into a particular problem defines a particular model. An example of this is the model of the hydrogen atom which is projected out of the Schrödinger equation of quantum mechanics by substituting the potential energy symbol in this equation with the potential energy of a central force field. The model then, is an emanation of theory and problem.

A phenomenon of modern times is "la ménagerie des modèles". It is an illness that affects mainly young disciplines that have not had the time to develop their own germinal concepts. Without support of a theory and with problems often ill-defined, there is a rush to the computer to construct "models" which, at best, generate trivial results. The illness has its clear origin in the aberrant deformation of putting the tool prior to the language, or the craftsmanship prior to the art.

1.2   ...Ecosystem ecology

Likens (1992) proposes "Ecology is the scientific study of the processes influencing the distribution and abundance of organisms, the interactions among organisms and the transformation and flux of energy and matter". Ecosystem ecology is the subdiscipline that is concerned with the transformation, flux and accumulation of energy and matter.

1.2.1 *The problems*

Ecology is a subject with many facets. In our endeavours to understand it as completely as possible, we need to make choices of perspectives; all questions are not viewed equally well from the same viewpoint. However, from any particular viewpoint we cannot see the whole system. Thus, by picking a fixed position from which to study the system, we lose information but hopefully what we see becomes more intelligible. The choice of viewpoint is, of course, critical. If we make the wrong choice, the picture we see is blurred and without any clear lines. There are no correct positions, only useful ones. When we happen to find ourselves facing any of these, our picture of the system may become clear and it will be possible to discern patterns and regularities in the system. The situation is perfectly analogous to viewing a solid, regular three-dimensional body. When viewed from an arbitrary angle it may look very complicated, but when examined along any of its principal axes it may appear as a simple geometrical shape, although we can now only see a limited portion of it. However, in a young science like ecology, it is not self-evident that an indistinct picture is a result of an unsuitable approach, it might just as well be due to our lack of insight in the area. We might need a pair of glasses (i.e. a theory) in order to see more clearly.

Figure 1.1 illustrates a possibility of projecting ECOLOGY into two different subspaces, an evolutionary, E, and an ecosystem, e, subspace. These subspaces are of different sizes and to some extent overlapping. When one subspace or part of it is contained in another, the laws in the smaller one must be subordinated to those in the larger subspace; the larger subspace is constraining the smaller one. The ecosystem subspace is often seen as a subspace of the evolutionary one. Laws at the ecosystem level may therefore not conflict with the evolutionary level. However, the ecosystem will also constrain evolution and it is not possible to point out any of the two subspaces as the more fundamental.

The choice of viewpoint also depends on the problems to be addressed. There are many relevant questions to be answered at the ecosystem level proper and where recourse to evolutionary principles leads to hopeless impasses. Similarly, there are questions that are meaningless at the ecosystem level but are key issues in the evolutionary perspective. In a more mature science such as physics, a comparable distinction can made between classical

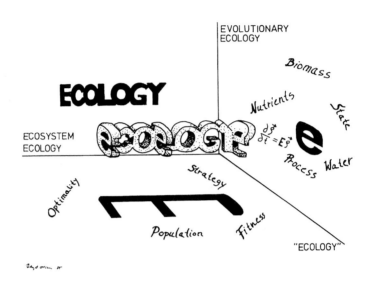

Figure 1.1 ECOLOGY projected onto the subspaces EVOLUTIONARY ECOLOGY and ECOSYSTEM ECOLOGY. The projections are made by looking along the axes labelled EVOLUTIONARY ECOLOGY and ECOSYSTEM ECOLOGY, respectively.

mechanics and quantum mechanics. It is quite clear that the student who tries using (relativistic?) quantum mechanics rather than classical mechanics to calculate the trajectory of his lecture notes for throwing them at his teacher will most likely not only flunk the exam but also miss his target, even though he/she was applying the most advanced and fully-fledged theory in the field. This is the perspective that should be kept in mind. Whatever we are doing at the ecosystem level should not conflict with evolutionary ecology, but it is not necessary that ecosystem statements can be derived from evolutionary ecology.

An important aspect when selecting the viewpoint from which one wants to observe, is the characteristic time constants thereby selected.[5] Any ecological system contains time constants, all of which are ecologically significant, ranging over many orders of magnitude. From any given viewpoint we should only try to look through a narrow window of time constants. The window chosen will, therefore, frame a few of them, which will then determine which processes are observable and which are not. Processes characterised by much longer time constants than the selected window evolve so slowly that they will not appear as processes at all but will seem to reflect some static properties of the system. Processes characterised by much shorter time constants will cycle through their entire spectrum of states and will only be represented by their average properties. Only those processes with time constants within the window give us a dynamic impression. The selection of a window therefore seriously restricts what we can understand about a system.

Although the narrowing of our views on a system precludes many statements and observations about the system in question, it is nevertheless a necessary restriction, if we are to avoid drowning. In technical systems it is not unusual to have to operate with subsystems of largely different time constants. These are commonly formulated as systems of stiff differential equations. When these cannot be solved analytically, which normally is the case, the problems involved in the numerical solutions may become formidable. The alternative to the exact solution is to proceed through different approximations. In physics, such approximations have been used for a long time and go under the name of adiabatic approximations. A typical case is to consider the atomic nuclei as fixed in space when calculating the dynamics of electrons. Because the typical difference in mass between an atomic nucleus and an electron is around 1800, and time constants are proportional to the square root of the mass, the time constants for motion of electrons and nuclei differ roughly by a factor of forty. The virtue of the adiabatic approximations lies not so much in the increased possibilities of solving the equations describing the motions, as in the increased understanding of the properties of the system emanating from the decoupling of the system into two essentially independent subsystems. Approximations or changes of view that allow decomposition of a complex problem into a series of simpler ones are indicators of fruitful approaches to the problem and should be explored in depth.

1.3   Setting a perspective

In this treatise, we see the ecosystem as the unit to be studied. Moreover, the perspective we want to apply is one where events in ecosystems are expressed and interpreted in terms of carbon and other elements.[6] In the next chapter, we will argue why this is one of the principal axes allowing us to get a clear view of ecosystems. The central concept linking together the dynamics of the elements in the ecosystem is "quality" which we denote with the symbol q; its meaning and properties are the subject of a large part of this book. The mathematical approach that we advocate can be summarised in an extremely general form

$$\frac{\partial \vec{\rho}(q,t)}{\partial t} = \mathbf{E}(q,t)\vec{\rho}(q,t) \tag{1.1}$$

where $\vec{\rho}(q,t)$ is the ecosystem state vector describing the distribution of carbon and elements in the ecosystem and $\mathbf{E}(q,t)$ the ecosystem operator. Equation (1.1) can be seen as an infinite set of models, and this will be discussed in greater detail in Chapter 11. In Chapters 3-12, we develop in depth the elements of the ecosystem state vector and ecosystem operator and use models projected from them to analyse a series of particular problems.

NOTES

[1] It is tempting to see this book as a refutation of Peters' book (1991) *A Critique for Ecology*. We will develop a theory and show how it goes far beyond simple regressions in explaining a large class of ecological phenomena.

[2] Bergner (199X) discusses extensively the meaning and interrelation between the words language, tool, theory, and model.

[3] The need to use mathematics in any mature science has been described elegantly by Bunge (1967, Chap. 8.2).

[4] An example of a very simple mathematical formula with a very complicated dynamics is the logistic difference equation which has been discussed, among others, by May (1976).

[5] A way of seeing the relation between different time constants is discussed at length by Allen and Starr (1982) and O'Neill et al. (1986). They suggest that systems with different time constants should be viewed as a hierarchy and their point is that longer time constants constrain shorter ones, i.e. the more rapidly evolving processes can proceed at the rate they do but only to the extent allowed by the slower processes. A typical example would be the variation in photosynthesis with time. The diurnal variations are large, yet what controls the plant's accumulation of carbon is not these very dynamic processes but rather the more slowly evolving, yet phenologically determined, development of leaf biomass. Another example is the growth of annuals within a matrix of perennials.

[6] Naturally, our choice of perspective is only one of many possible ones. H.T. and E.P. Odum have emphasised energy as the natural currency in ecosystem research (Odum 1971ab, Odum 1983). Our views should not be seen as necessarily conflicting with theirs, but as alternatives. There is also a difference in scope. We restrict ourselves to the ecosystem as such, whereas, in particular, H.T. Odum wants his approach to encompass human society as well. A fundamental of ecological systems is, in a thermodynamic sense, their open character. Theoretically, this has far-reaching implications although at present its practical consequences have not been demonstrated. Experience obtained from the much simpler physical systems tells us that open systems are full of surprises (Glansdorff & Prigogine 1971; Haken 1977). It is also obvious that the search' for extremal principles and equations of state is going to be very difficult, cf. the difference in going from free energy to entropy production. It is important that (1.1) is linear because most non-linear equations are insoluble analytically (Constanza et al. 1993). A consequence is, however, that certain classes of problems cannot be described; see also Epilogue.

2

Element cycling

Element cycling is the term used to describe the movement of certain materials into, within, and out of an ecosystem. This dynamics is possible because there are sufficient supplies of energy in the ecosystem. In this book, the materials cycled are the elements of the periodic table. We discuss in this chapter why we have decided to concentrate on the elements carbon and nitrogen.

2.1   Elemental distribution

Because the perspective of this text is one where the living world is largely viewed as a chemical machine, it may be worthwhile looking at the chemistry of living organisms in relation to their environment. A summary of the chemical composition of a plant, an animal, and the earth is given in Table 2.1.

Striking features in Table 2.1 are the great similarities between different organisms in their chemical composition and their dissimilarity to their chemical environment. A few elements like Ca and Na differ by a factor of three between man and alfalfa but C and N differ by a factor exceeding 500 between the living and the inanimate. The only element that appears in approximately the same proportion everywhere is oxygen. However, the major part of the oxygen in living organisms is bound in water and when water is excluded the organism's ability to enrich in the few elements within the box in the Table becomes even more apparent. This emphasises the necessity to regard living organisms as open systems. It is also important to recognise that when dealing with the living and its chemical environment, we frequently encounter systems at steady state but far from chemical equilibrium, i.e. systems in which the fluxes in and out are in balance but where a continuous energy expenditure is required to maintain these fluxes.[1]

Table 2.1 *Distribution of elements in the earth's crust, a plant (alfalfa), and an animal (man) in per cent by weight (Handbook of Chemistry and Physics 1975; Bertrand 1950).*

Element	Earth's crust	Plant	Animal
O	47.0	78.0	65.0
Si	28.0	0.009	0.004
Al	8.0	0.003	0.0
Fe	5.0	0.003	0.005
Ca	3.6	0.6	2.0
Na	2.8	0.03	0.1
K	2.6	0.2	0.2
Mg	2.1	0.08	0.04
H	0.14	9.0	10.0
P	0.12	0.7	1.0
S	0.052	0.1	0.6
C	0.032	11.0	18.0
N	0.0046	0.8	3.0

2.2   Carbon and nitrogen

The time has now come to argue why we should concentrate on nitrogen and carbon. Any treatise on ecology must include carbon as an element for the obvious reason that carbon is what defines the organic system. Other elements to be included are less obvious. Here we have mainly taken nitrogen, phosphorus and sulphur into consideration. These elements have one important feature distinguishing them from all the other elements; essentially all the reservoirs involved in biological circulation are in organic form - the stores of inorganic forms accessible to living organism are generally very small relative to organically bound forms. Hence, any circulation of these elements is intimately coupled to biological processes. This is in contrast to other elements which can be stored in large quantities in inorganic form in the soil. The understanding of the processes involving elements other than nitrogen, phosphorus, sulphur and carbon is, therefore, basically one of inorganic chemistry where basic principles are known. The understanding of many of the biological processes in the transformations of C, N, P and S lacks theoretical foundations and there are few basic principles on which one can rely. It is this void that we hope to fill to some extent. We will in many cases argue that nitrogen, phosphorus, and sulphur act in ecologically similar ways and that the same formalism can be applied to any of them, albeit with numerically different parameters. Most of our concern is, however, with nitrogen as this is generally considered to be the element limiting plant growth in most instances.

The global cycles of carbon and nitrogen are characterised by enormous storages in the pedosphere, the turnovers of which are virtually zero and their contributions to the cycles are therefore small. Another feature of the distributions of these elements is the large pools found in the oceans, and in the case of nitrogen in the atmosphere in the form of N_2.

We have summarised some gross characteristics of the terrestrial part of the global cycles of carbon and nitrogen in Table 2.2.[2] Some of the features worth noticing are how dominating the soil storage of nitrogen is relative to carbon. As a consequence, the turnover of soil nitrogen is slow in relation to the turnover of plant nitrogen. However, this is partly an effect of how plants are described, as there is considerable internal cycling within the plants. The relative scarcity of nitrogen in plants and the relative scarcity of carbon in soil indicate that the rates of turnover may be limited by nitrogen for plants and carbon for soil. In fact, our subsequent discussions will be based on the assumption that plant production is nitrogen-limited and soil organic matter turnover is carbon-limited.[3]

From physical and biochemical points of view it may be argued that the acquisition rate of carbon should not be limiting but rather the rate at which the primary photosynthetic products in the plant could be turned into more complex compounds, where the latter rate is determined by availability to nitrogen.[5] The argument is as follows. Consider a leaf with a typical boundary layer of 1 mm and an external CO_2 concentration of 350 ppm and zero inside.[6] The use of an internal concentration of zero rather than some finite value is the extreme a plant can achieve through the allocation of its resources. Since the diffusion rate of CO_2 is 0.14 $cm^2 s^{-1}$, this gives a maximum assimilation rate of 8 g C $(m^2$ leaf area$)^{-1} h^{-1}$. Typical specific leaf areas are 20-25 $m^2 kg^{-1}$, leading to relative growth rates of 35-40% h^{-1}, which is well above any observation of plant growth.

Table 2.2 *Gross characteristics of the terrestrial part of the global carbon and nitrogen cycles. Pg = 10^{15} g (Adapted from Ajtay et al. 1979 and Bolin 1979).*[4]

		C	N	C/N
Storage, Pg	Plants	560	12	47
	Soil	1600	300	5
Ratio Soil/Plants		3	25	
Flow, Pg/yr	Plants-to-Soil	43	2	21
Turnover times, yr	Plants	13	6	
	Soil	37	150	

Almost all the nitrogen in green plant material is bound in proteins.[7] Evolution should have created a balance between different proteins, the activity of any one protein can therefore be used to estimate the efficiency of all. Of the proteins, D-ribulose-1,5-biphosphate carboxylate/oxygenase (Rubisco) is by far the dominating one, constituting 15-30% of all leaf proteins. Using Rubisco as a model protein, it can be estimated that per unit of nitrogen, 0.9 g C can be fixed per hour. Assuming 1 g N $(m^2$ leaf area$)^{-1}$ gives the production rate of 0.9 g C $(m^2$ leaf area$)^{-1}h^{-1}$, which is an order of magnitude lower than the limitation calculated from carbon dioxide uptake. This production rate is comparable to what is required of plants growing at extremely high growth rates,[8] e.g. relative growth rates of 100% d^{-1} and nitrogen concentrations of 0.10 g N $(g\ C)^{-1}$ giving 0.4 g C $(g\ N)^{-1}h^{-1}$. This suggests that the total nitrogen amount in a plant can be used to calculate its production rate. We will use the term *nitrogen productivity* to denote the amount of dry matter (or carbon) produced per unit of nitrogen and unit of time. The whole of Section III is devoted to demonstrating how the nitrogen productivity can be used to analyse plant growth.

NOTES

[1] One striking example of a system at steady state but far from chemical equilibrium is N_2 in the biosphere. At present oxygen concentrations and pH

in the oceans, practically all N_2 in the air ought be converted to nitrate for a true equilibrium to be established (Sillén 1966).

[2] The occurrence of soil organic phosphorus was reviewed by Dalal (1977) and he concluded that organic phosphorus constitutes 20 to 80% of the total phosphorus in the surface layer of the soil. The importance of the organic phosphorus can be greater because in the soil water solution around 80% of the phosphorus is organic but with some variation depending on soil type (Pierre & Parker 1927). Regarding sulphur, Biederbeck (1978) states that "well over 90% of the total S in most noncalcareous surface soils is present in organic form". Scott (1985) goes even further: "It is well established that 95% or more of the total sulphur in most soils from humid and semihumid areas is organic" The C:S (as the C:N:S) ratio is also restricted within rather narrow limits, between 70 and 300 (Biederbeck 1978; Stewart 1984).

[3] Schlesinger (1991) argues that nitrogen should be the most commonly limiting element because it is the only element (disregarding carbon) that occurs in forms (NO_3^-, NO_x, and N_2) that readily can be lost from a system.

[4] The values given in Table 2.2 are adapted from Ajtay et al. (1979) for carbon and from Bolin (1979) for nitrogen. These values must be treated with some reservation as they are subject to constant revision. For example, a more recent estimate of the storage of nitrogen in soils is 95 Pg (Post et al. 1985).

[5] For arguments concerning growth limitation in plants, see Ågren (1985b)

[6] The internal CO_2 concentration in plants was measured to 200 ppm by Jarvis & Sandford (1986).

[7] Mengel & Kirkby (1979) give a value of 85% for the fraction of nitrogen in green plant material bound to proteins.

[8] The value 0.4 $gC(gN)^{-1} h^{-1}$ for the production rate is comparable to the potential photosynthetic nitrogen-use efficiencies of 0.1-0.4 $gC(gN)^{-1}h^{-1}$ calculated by Field & Mooney (1986) for a number of wild plants.

PART II

THE SOIL

- Mums, vilket Q, sa Q-hunden.
Fina Q! Förlåt mig, men jag kunde inte låta bli.
Ett stort Q, runt och mulligt med en liten knaprig svans på.
Härligt, härligt!

(H. Alfredson *Varför är det så ont om Q*)

- Yummie, what a Q, said the Q-dog.
Fine Q! Forgive me, but I could not help it.
A big Q, round and chubby with a small crunchy tail.
Delicious, delicious!

(H. Alfredson *Why are there so few Qs*)
translated by the authors

3

Theory for homogeneous substrates

This chapter serves as an introduction to the formalisation of the carbon and nitrogen turnover in soils. Terminology and a number of concepts are introduced by considering a simple, idealised model. The main purpose is to derive equations of a general character that describe in an appropriate way the dynamics of carbon and nitrogen in decomposing litter structures. Emphasis is on the method: few variables and analytical solutions. We start with very general formulations; it is important to realise where different approximations are introduced and to understand clearly the meaning of such approximations. A basic concept that is going to be useful throughout is the concept of "litter cohort"; a litter cohort is a homogeneous quantity of litter that enters the soil at the same time. We assume that the soil organic matter consists of populations of litter cohorts and that properties emerge as sums or integrals over these populations. Significant similarities between the cycles of N, P and S in the soil have been identified in the literature. Many of the properties discussed for N can be extended also to the other elements. However, for simplicity here we restrict the analysis to N. In the next chapter and especially in Chapter 6, extensions to other elements are discussed in detail. In this chapter we deal with a type of organic matter that we call homogeneous and imply thereby that there is no need to separate the organic matter into different components; it can at all times be regarded as one homogeneous substrate.

3.1   Main concepts

Our basic assumption is that decomposition of organic matter is the main process behind the cycling of carbon and nitrogen in the organic matter of the soil, and the rate by which organic compounds are catabolised, i.e. transformed from organic to inorganic forms, is determined by the activity of a decomposer community.[1] The carbon dynamics of the decomposing litters shows a monotonously decaying behaviour while it is possible to distinguish distinct phases in the nitrogen dynamics of the same litters.[2] Figure 3.1 illustrates the phases of leaching, accumulation and release of nitrogen found in decomposing needle litter. Other kinds of litters may show different behaviours, starting directly at the accumulation or at the release phase. During the accumulation phase, the nitrogen in the decomposing material is in short supply with respect to the needs of the decomposers, which must import inorganic nitrogen from the surroundings to maintain the balance between the elements: the material is nitrogen deficient. On the other hand, during the release phase there is a surplus of nitrogen and, consequently the material is carbon or energy deficient.

The value of the N:C ratio[3] at the critical time t_c (Figure 3.1) viz., the value of the N:C ratio at which a net release of nitrogen is started, is called the *critical N:C ratio*. This concept originates in agricultural experiments on decomposition. Several authors[4] have shown that crop residues under decomposition immobilise available external nitrogen until they reach a N:C ratio of about 0.03 to 0.04 (1.4 - 1.7% N). If the added plant material already has a N:C ratio greater than this critical range, nitrogen is released as ammonia. The nitrogen factor,[5] a concept closely related to the critical N:C ratio, was defined many years ago as the amount of nitrogen accumulated in the organic material before a net release begins. Knowledge of this factor, viz., of the amount of nitrogen required to "decompose" a given residue was considered of main interest for the determination of usable soil amendments in agriculture. The nitrogen factor is defined as the difference between the maximum and the initial amounts of nitrogen.

3.2   Carbon and nitrogen dynamics

Let an amount $B(t)$ of living decomposer biomass be growing in a homogeneous substrate with a certain amount $C(t)$ of carbon and $N(t)$ of nitrogen, producing new biomass at a rate $P(t) > 0$. Let us further define the decomposers by the following properties: i) f_C (gC g^{-1}) and f_N (gN g^{-1}), the *carbon and nitrogen concentration*, respectively, in the decomposer biomass; ii) e_0 (dimensionless), the *decomposer efficiency* or the production-to-assimilation ratio of the decomposers feeding on the substrate; and iii) u (gB $gC^{-1}time^{-1}$) is the *decomposer growth rate per unit of carbon*. The efficiency with which the carbon of the substrate is utilised for decomposer growth also includes maintenance energy costs. At this point we assume that f_C, f_N, e_0, and

u are constant in time. We also assume that dead decomposer biomass is returned to the original substrate at a rate $M(t)$ with the same carbon and nitrogen concentrations as in B and with the same characteristics (other than concentrations) as the original substrate. This last assumption and that of the constancy of decomposer properties are compatible with the definition of a homogeneous substrate. In this ideal substrate, only the amounts of matter and nutrients can change in time.

Since our focus is on long-term dynamics, we assume that we can ignore the details of structure of the decomposer community. One of our methods for solving the equations describing the decomposition will be to assume that the decomposer biomass is in equilibrium with the available substrate. We will therefore assume that, to the extent that changes in the decomposer commu-

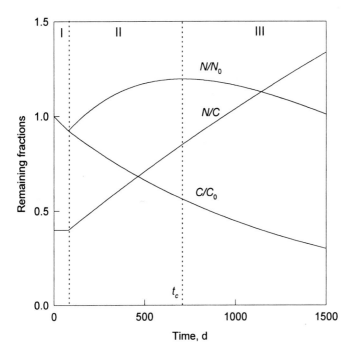

Figure 3.1 Remaining nitrogen and carbon (as fractions of absolute initial amounts N_0 and C_0) and N:C as a function of time in decomposing forest litter (after Berg & Staaf 1981). The three phases of leaching (I), immobilisation (II) and release (III) in the dynamics of nitrogen are shown, as well as the critical time, t_c, at which nitrogen reaches the critical concentration after which a net release of nitrogen is started from the litter.

nity occur, this can be dealt with by replacing decomposer parameters with functions of substrate properties.

Figure 3.2 gives a representation of the processes associated with the decomposition of a homogeneous substrate. Decomposers incorporate carbon and nitrogen into biomass at rates $f_C P$ and $f_N P$, respectively. According to the definition of e_0, $f_C P/e_0$ is the rate at which carbon is taken up from the substrate. The rate of respiration, R, is the difference between the rate of utilisation and uptake

$$R = \frac{f_C P}{e_0} - f_C P = f_C (\frac{1}{e_0} - 1) P \qquad (3.1)$$

R is the rate at which total carbon, i.e. substrate carbon plus decomposer carbon, is lost from the system. In a similar way, the rate of nitrogen release, m, is defined as the difference between the rate of nitrogen uptake and the rate at which it is incorporated into biomass. There is no factor comparable to e_0 for nitrogen. Instead, we assume that nitrogen utilisation passively follows

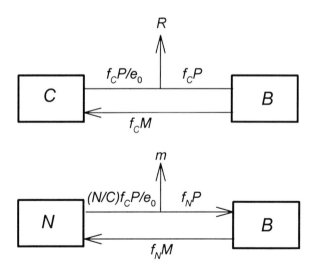

Figure 3.2 Diagram of the processes related to the decomposition of a homogeneous substrate. C and N are absolute amounts (e.g. g m^{-2}) of carbon and nitrogen. B is decomposer biomass and P and M are the rates of decomposer production and mortality, respectively.

carbon utilisation. Thus, nitrogen uptake is equal to the product of carbon uptake and the substrate N:C ratio

$$m = \frac{f_C}{e_0} P \frac{N}{C} - f_N P \tag{3.2}$$

or

$$m = \left(\frac{r}{r_c} - 1\right) f_N P \tag{3.3}$$

where

$$r = N/C \tag{3.4}$$

is the N:C ratio of the substrate and

$$r_c = \frac{e_0 f_N}{f_C} \tag{3.5}$$

Since $P > 0$ and $0 < e_0 < 1$, R is always positive (3.1) and thus, the total amount of matter decreases monotonously with time. On the other hand, m can be either positive or negative depending on the amount of nitrogen in the substrate (3.2). For this reason, we identify m with the rate of net nitrogen mineralisation: when m is positive, nitrogen is released from the substrate, but when m is negative nitrogen is accumulated. Note, however, that we are not (see however Chapter 7) defining m as the difference between gross mineralisation and immobilisation rates but, as the difference between nitrogen uptake and decomposer requirements (3.2); if there is sufficient nitrogen in the substrate to supply the decomposers, nitrogen is released (mineralised), otherwise decomposers import (immobilise) inorganic nitrogen from the surroundings.

When the substrate is carbon-deficient, $r > r_c$ and $m > 0$ (3.3), and nitrogen is mineralised. On the other hand, for a nitrogen-deficient substrate $r < r_c$, and $m < 0$ and nitrogen is immobilised. This means that r_c can be identified with the critical N:C ratio.[6]

The simple conceptual scheme in Figure 3.2 gives a good representation of the behaviour of nitrogen mineralisation near the critical point. In order to analyse this behaviour for arbitrary times we must obtain the dynamic equations for C, N and B.[7] These are (Figure 3.2):

$$\frac{dC}{dt} = -\frac{f_C P}{e_0} + f_C M \tag{3.6a}$$

$$\frac{dB}{dt} = P - M \tag{3.6b}$$

$$\frac{dN}{dt} = -\frac{f_C P}{e_0}\frac{N}{C} + f_N M \tag{3.6c}$$

The mortality function can be expressed in terms of a lag time (mean life time of the decomposers), τ ($\tau > 0$), in the productivity function P, i.e.

$$M(t) = P(t - \tau) \tag{3.7}$$

$P(t - \tau)$ can be expanded in a Taylor series in τ,

$$P(t - \tau) = \sum_{n=0}^{\infty} \frac{(-\tau)^n}{n!} \frac{d^n P}{dt^n} \tag{3.8}$$

Equation (3.8) is an expansion of the mortality function in all orders of τ. Taking the zeroth order approximation, $P(t - \tau) \sim P(t)$, (3.6) transforms into (see section 3.4 for higher order approximations)

$$\frac{dB}{dt} = 0 \tag{3.9a}$$

$$\frac{dC}{dt} = -\frac{1 - e_0}{e_0} f_C P \tag{3.9b}$$

$$\frac{dN}{dt} = -\left(\frac{r}{r_c} - 1\right) f_N P \tag{3.9c}$$

Neglecting all lag-time effects is therefore equivalent to assuming that the decomposer biomass is in a steady state condition (adiabatic approximation) or, equivalently, that the dynamics of C and N in the decomposer biomass are indistinguishable from the dynamics of the total amounts of carbon and nitrogen in the substrate.[8] A more accurate condition to validate the use of the zeroth order approximation is derived in section 3.4.

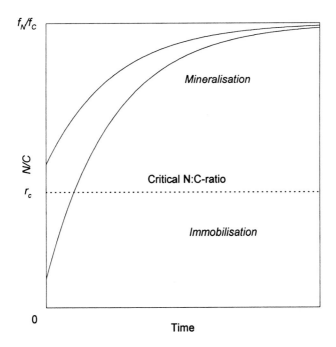

Figure 3.3 The temporal development of the N:C ratio for a homogeneous substrate. If the initial N:C ratio, r_0, is smaller than the critical value, $r_c = e_0 f_N/f_C$, nitrogen will first be immobilised. If r_0 is greater than r_c, mineralisation occurs all the time. In both cases, the final value of the N:C ratio in the substrate is the ratio between the nitrogen and carbon concentrations, f_N/f_C, in the decomposer biomass. The same pattern of behaviour is obtained for any functional choice of decomposer productivity.

A series of results can be deduced from (3.9) without making any additional assumptions whatsoever about the nature of the decomposer production P. We have already shown that C decreases monotonously in time. To find the behaviour of N, we first take the time derivative of the N:C ratio, r, and use (3.9b) and (3.9c) to obtain

$$\frac{dr}{dt} = \left(\frac{f_N}{f_C} - r \right) \frac{f_C P}{C} \qquad (3.10)$$

Let C_0, N_0 be the amounts of C, N and $r_0 = N_0/C_0$ be the N:C ratio at time $t = 0$. Since P/C is a positive quantity, r will, according to (3.10), increase ($dr/dt > 0$) or decrease ($dr/dt < 0$) with time depending on whether $r_0 < (f_N/f_C)$

or $r_0 > (f_N/f_C)$ until it reaches the final value f_N/f_C. Two situations are of practical interest. Assume first that the substrate is initially carbon-deficient, i.e. $(f_N/f_C) > r_0 > r_c$ (upper curve in Figure 3.3). Under these conditions, $m > 0$ all the time and N will decay steadily in time. The other situation is $(f_N/f_C) > r_c > r_0$, i.e. the substrate is nitrogen-deficient (lower curve in Figure 3.3). From (3.3), $m < 0$ and N will increase (3.9c) from its initial value N_0. But from (3.10) r increases with time and when it reaches the value r_c, m equals zero and N reaches a maximum (3.9c). Thereafter, the substrate becomes C-deficient and a net release of N begins. The conclusion is that, regardless of the definition adopted for P, (3.9b) and (3.9c) predict the accumulation and release phases of nitrogen in the decomposing substrate.

Problem 3.1
Derive equation 3.10

3.3   A particular solution

In order to deduce more specific results, P must be defined in some explicit way. Let us assume that[9]

$$P = uC \tag{3.11}$$

P proportional to C implies that the growth of decomposers is limited by the energy sources and not by N. During the immobilisation phase, when nitrogen is in short supply in the substrate, we assume that decomposers can utilise inorganic sources of N in the soil.

Introducing (3.11) into (3.9b) and (3.9c) we get[10]

$$\frac{dC}{dt} = -kC \tag{3.12a}$$

$$\frac{dN}{dt} = -\frac{f_C u}{e_0} N + f_N uC \tag{3.12b}$$

where

$$k = \frac{1 - e_0}{e_0} f_C u \tag{3.13}$$

Equation (3.13) explicitly relates the specific decomposition rate to decomposer properties.

Equation (3.12a) can be integrated giving

$$C(t) = C_0 e^{-\frac{1-e_0}{e_0} f_C u t} \tag{3.14}$$

and with (3.14), (3.12b) can be integrated. Since $P/C = u =$ constant, (3.11), it is easier to integrate (3.10), which gives[11]

$$r(t) = \frac{f_N}{f_C} - \left(\frac{f_N}{f_C} - r_0\right) e^{-f_C u t} \tag{3.15}$$

showing that, when $t \to \infty$, r approaches the asymptotic value f_N/f_C.

For future use we want to introduce the functions $g(t)$ and $h_N(t)$ which define, respectively, the amounts of C and N in the substrate with respect to the initial amounts of C, i.e.

$$g(t) = \frac{C(t)}{C_0} \tag{3.16}$$

$$h_N(t) = \frac{N(t)}{C_0} = r(t)g(t) \tag{3.17}$$

Figure 3.4 shows the dynamic behaviour of these variables.

At the critical time, t_c, (3.15) must be equal to (3.5). Imposing this condition, we derive the following equation

$$t_c = -\frac{1}{f_C u} \ln \frac{1-e_0}{1 - \dfrac{r_0}{f_N / f_C}} \tag{3.18}$$

In general $r_0 < (f_N/f_C)$, decomposer biomass has a higher N:C ratio than plant litter, and the term in the logarithm function is therefore a positive quantity. If also $r_0 < (e_0 f_N/f_C)$, then this term is less than 1 and t_c is positive, indicating the presence of an accumulation phase. Combining (3.14), (3.15), (3.17), and (3.18) we derive the equation for h_c, the maximum value of $h_N(t)$

$$h_c = \frac{e_0 f_N}{f_C} \left[\frac{1-e_0}{1 - r_0/(f_N / f_C)}\right]^{\frac{1-e_0}{e_0}} \tag{3.19}$$

Equation (3.19) is, then, a definition of the nitrogen factor in terms of decomposer and plant litter properties.

Problem 3.2

Verify that (3.15) satisfies (3.10)

Problem 3.3

Derive (3.18) and (3.19)

Problem 3.4

Analyse the effects of different parameter values upon $g(t)$ and $h_N(t)$ in Figure 3.4.

Problem 3.5

Derive (3.38), (3.39a) and (3.39b).

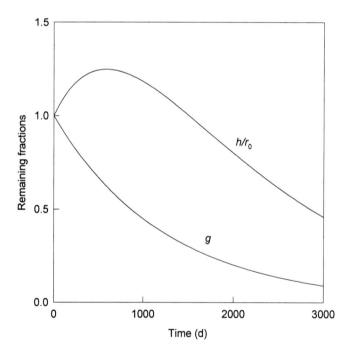

Figure 3.4 The phases of accumulation and release in the dynamics of N as predicted with equations (3.16) and (3.17). Carbon decays exponentially to zero. The curves are calculated with the following values of the parameters: $e_0 = 0.2$, $f_C = 0.5$, $f_N = 0.05$, $u = 0.0004 \text{ d}^{-1}$, $r_0 = 0.01$.

3.4   Lag-time effects

We shall now analyse the implications of including lag times up to the first order in the previous equations. Let us assume that we inocculate the substrate with an amount of decomposer biomass corresponding to its equilibrium with the substrate at $t = 0$. Then $P(t) = 0$ for $t < 0$ and from (3.7) we have for all times $t < \tau$

$$M(t) = P(t - \tau) = 0 \qquad (3.20)$$

Then, with the definition (3.11) combined with (3.9b) and (3.9c) we get, for $t < \tau$

$$\frac{dC}{dt} = -\frac{f_c u}{e_0} C \qquad (3.21a)$$

$$\frac{dN}{dt} = -\frac{f_c u}{e_0} N \qquad (3.21b)$$

i.e. C and N follow negative exponentials with the same specific decay rate.

For times $t > \tau$ we take the series expansion (3.8) up to terms of first order in τ, i.e.

$$P(t - \tau) \sim P(t) - \tau dP/dt \qquad (3.22)$$

or, using (3.11)

$$M(t) = P(t - \tau) \sim u[C(t) - \tau dC/dt] \qquad (3.23)$$

Introducing (3.23) into (3.6a) and integrating we get

$$C(t) = C(\tau)e^{-k_1(t-\tau)} \qquad (3.24)$$

where

$$k_1 = \frac{1 - e_0}{e_0} \frac{f_c u}{1 + f_c u \tau} = \frac{k}{1 + f_c u \tau} \qquad (3.25)$$

and $C(\tau)$ is the solution to (3.21a) taken at time τ, i.e.

$$C(\tau) = C_0 e^{-f_c u \tau / e_0} \qquad (3.26)$$

Introducing (3.11) and (3.23) into (3.6c) and using (3.24) we get the dynamic equation for $N(t > \tau)$

$$\frac{dN}{dt} = -\frac{f_C u}{e_0} N(t) + f_N u(1 + k_1 \tau)C(t) \tag{3.27}$$

This equation has the solution $(t > \tau)$

$$N(t) = N(\tau)e^{-k_2(t-\tau)} + \frac{f_N u(1 + k_1 \tau)}{k_2 - k_1} \left[1 - e^{(k_1 - k_2)(t-\tau)}\right] C(t) \tag{3.28}$$

where $k_2 = f_C u/e_0$ and

$$N(\tau) = N_0 e^{-k_2 \tau} \tag{3.29}$$

is the solution to (3.21b) taken at time τ.

Figure 3.1 shows the behaviour of C (3.24) and N as given by the solution to (3.21b) and (3.28) with a lagtime of 80 d and other parameters as in Figure 3.4. Since for $t < \tau$, C and N decay at the same rate (equations 3.21), the N:C ratio remains constant over this period. If the product $u\tau$ is small, i.e. if $u\tau \ll 1$, then the equations (3.21), (3.24) and (3.28) of the first order approximation differ very little from the zeroth order approximation, (3.14) and (3.15). In other words, for soils, in which the condition $u\tau \ll 1$ applies, the zeroth order approximation can safely be used in formulating the dynamics of carbon and nitrogen.

Problem 3.6
Derive (3.24) and (3.27).

Problem 3.7
Verify that (3.28) satisfies (3.27).

3.5   From single litter cohorts to the soil

Let us consider a soil that has been receiving a continuous input of litter for a number of years. Denote with $I(t)$ the rate of input in units of carbon (e.g. $gC\ m^{-2}yr^{-1}$), where $I(t)$ is a general function of time. The amount of carbon in the litter cohort that entered in the small interval dt' is $I(t')dt'$ and, by the definition of g, (3.16), the amount of carbon remaining at time t $(t > t')$ is $I(t')g(t - t')dt'$. Let $a = t - t'$ denote the age of a litter cohort. Since the first cohort entered the soil at $t' = 0$, t is the age of the oldest cohort and $I(t - a)da$ is the initial amount of carbon in the litter cohort of age a. We assume now that the soil consists of a mixture of non-interacting litter cohorts, each characterised by its age a and each decomposing according to $g(a)$ and $h_N(a)$. In particular, the amounts of carbon and nitrogen in the soil, $C_s(t)$ and $N_s(t)$, can be calculated by summing the contributions of all cohorts, i.e.

$$C_s(t) = \int_0^t I(t-a)g(a)da \qquad (3.30)$$

$$N_s(t) = \int_0^t I(t-a)h_N(a)da \qquad (3.31)$$

Suppose that the rate of input is constant, I_0, and that $g(a)$ is defined by (3.14) and (3.16), then

$$g(a) = e^{-ka} \qquad (3.32)$$

where k is given in (3.13). Evaluating the integral in (3.30), we get

$$C_s(t) = \frac{I_0}{k}\left(1 - e^{-kt}\right) \qquad (3.33)$$

and taking the derivative of (3.33) with respect to time

$$\frac{dC_s}{dt} = I_0 - kC_s \qquad (3.34)$$

According to (3.33), $C_s(0) = 0$; as more and more litter cohorts enter the soil, $C_s(t)$ accumulates until a steady state (ss) is reached, at which $dC_s/dt = 0$ and[12]

$$C_s^{ss} = \frac{I_0}{k} \qquad (3.35)$$

More explicitly, in our case we get, using (3.13)

$$C_s^{ss} = \frac{e_0}{1-e_0}\frac{I_0}{f_C u} \qquad (3.36)$$

which relates the steady state value to decomposer properties.

In the next chapter we will treat the situation where the decomposability of a litter cohort changes with age.

Problem 3.8

Use (3.1) together with (3.11) and (3.14) to calculate the rate of mineralisation of carbon (respiration) from the whole soil. Calculate this value at ss.

Problem 3.9

Use (3.14) and (3.15) in order to calculate $N_s(t)$. Calculate N_s^{ss} and the rate of mineralisation of nitrogen from the soil at ss.

Problem 3.10

Calculate $r^{ss} = N_s^{ss}/C_s^{ss}$ and compare it with the critical value r_c. Is the ss of the soil C or N limited?

NOTES

[1] Heal (1979) has described the decomposition subsystem as a complex interacting set of organisms and processes, whose activities are central to the availability of nutrients to plants and in the maintenance of the nutrient capital of the ecosystem.

[2] Basic features of nitrogen and carbon transformations can be identified from litter bag studies of whole litter structures. For example, both weight loss and the N:C (nitrogen to carbon) ratio change in forest litters are quite regular and a strong positive relationship exists between nitrogen concentration and weight loss (e.g. Howard & Howard 1974; Aber & Melillo, 1980; Berg & Staaf 1980, 1981).

[3] We will use the N:C ratio rather than the conventional C:N ratio as the former has a number of advantages from a theoretical point of view as will be obvious from our subsequent analyses.

[4] Already in 1927, Waksman & Tenney attempted a qualitative explanation of the accumulation and release phases of nitrogen, see also Richards & Norman (1931) and Allison (1973). Swift et al. (1979) proposed that the same general trends of the accumulation-release dynamics should also apply to other nutrient elements than nitrogen, and they claimed that the carbon-to-element ratio in the resource and in the decomposer biomass are the main factors determining the presence of, or the lack of, an accumulation phase of the nutrient in the material.

[5] See Richards & Norman (1931).

[6] In 1975, Parnas postulated a mathematical definition of the critical N:C ratio by relating it to three decomposer properties; we have now derived the same relation, equation (3.5).

[7] Zheng et al. (1997) have shown how detailed descriptions of the decomposer community can be aggregated to an ecosystem level.

[8] An alternative way of deriving (3.9) is by assuming that the decomposer biomass is a small fraction of the total organic matter in the soil, a condition found in most natural ecosystems (Bosatta & Berendse 1984; Anderson & Domsch 1989).

[9] The proportionality between decomposer growth rate and substrate amount was used by Bosatta & Staaf (1982) and Bosatta & Berendse (1984).

[10] Equation (3.12a) coincides with the model for decomposition first suggested by Olson in 1963, namely a negative exponential with a constant mass loss rate.

[11] For a number of forest plant residues, nitrogen concentration has been linearly related to weight losses (Aber & Melillo 1980) or, in terms of r and g

$$g = A - Br \qquad (3.37)$$

where A and B are positive quantities. Using (3.14) to eliminate the exponential term in (3.15), we get

$$g = \left[\frac{f_N / f_C - r}{f_N / f_C - r_0} \right]^{\frac{1 - e_0}{e_0}} \qquad (3.38)$$

If decomposition has not proceeded too far (first years), we can assume that $r \sim r_0 \ll (f_N/f_C)$. Expanding (3.38) in a McLaurin's series, we get equation (3.37) with

$$A = 1 + Br_0 \qquad (3.39a)$$

$$B = \frac{1 - e_0}{e_0 (f_N / f_C - r_0)} \qquad (3.39b)$$

With the values given to the parameters in Figure 3.4 we get $A \sim 1.6$ and $B \sim 57$, which is within the range of values found by Aber & Melillo.

[12] (3.35) is the expression normally used in the literature (see e.g. Swift et al. 1979).

4

Theory for heterogeneous substrates

In Chapter 3 we considered the homogeneous substrate where only the amounts of the elements change over time. In this chapter,[1] the consequences of changing substrate properties during the decomposition process are taken up by introduction of the fundamental variable *substrate quality* - a measure of substrate accessibility to degradation. The conceptual idea is that a given chemical substance in the substrate can be associated with a specific quality; in this way the litter cohort, the organic matter, and the soil can all be seen as heterogeneous substrates composed of a mixture of qualities. Even if, initially, it were possible to identify a substrate with a single value of quality, an array of interactions - physical, chemical and biological - will at later times, produce a *dispersion* in quality (the single quality is spread into a distribution of qualities). Dispersion in quality becomes, then, a measure of substrate heterogeneity.

We first formulate the theory of decomposition and derive a set of general models depending upon the two previously introduced decomposer functions, viz., the decomposer *efficiency* of substrate utilisation, the decomposer *growth rate* and a new one, the decomposer *dispersion*. The last function describes how the quality of the assimilated carbon is distributed when it is returned to the substrate and is the force driving the change in quality. At this point, the formulation is general enough to allow even processes not mediated by decomposers but transforming quality at a given rate and with a given efficiency to be incorporated into the theory. The main dynamic variables are the *density function of carbon* - a measure of the probability of a C-atom being found in a given infinitesimal volume element of the quality space - and a similar nutrient density function for an element n, where n is any of N, P or S. The extension from N to phosphorus and sulphur is another important generalisation in this chapter.

We will specify some particular models for the three decomposer functions. In this way, it becomes possible to calculate $g(t)$ and $h_n(t)$ - the amounts of carbon and element n relative to the initial amount of carbon in the litter cohort - and make the consequent extensions to the organic matter and the soil which is the subject of forthcoming chapters. By representing the dynamics of C, N, P and S as functions of quality rather than time, aspects concerning mainly the rates of processes are projected out and the essentials of the decomposition process become more evident and easier to analyse.

4.1   Substrate quality

The cycling of C, N, P and S in the organic matter of the soil involves a complex interacting set of organisms and processes. During decomposition, organic matter in the soil is incessantly transformed between different chemical compounds. However, there is an overall trend towards more refractory components: fresh, easily decomposable, organic matter enters the decomposition systems and ends up as refractory, quasi-undecomposable, humic substances. We will use the term *quality* and the symbol q to summarise the attributes of the organic matter that determine this trend. We define quality as a measure of substrate accessibility expressed through the growth performance of the decomposer community; in formal terms $u(q) > 0$ and $du/dq > 0$.

Quality of litters and soil organic matter and its change with time has been investigated by several authors[2] to find which chemical substance or substances can be used to represent it. Soil organic matter, on the other hand, is rather divided into fractions of different ages and characterised by different turnover rates (qualities). Our purpose is a different one; we do not primarily want to find methods of associating specific chemical fractions with quality but rather to analyse the consequences of changing qualities, and as we show, many results can be obtained without such associations.[3] We will return to this question in depth in Chapter 11, where we will demonstrate one way of measuring quality in terms of chemical variables.

4.2   Basic equations for carbon

We start by deriving a system of general equations for carbon, nutrients, and decomposer biomass. The system is very general as it is only an expression of mass balances. Consider a substrate that is decomposed by a decomposer biomass. We introduce the continuous variable q $(0 < q < \infty)$ for quality and the density function of carbon $\rho_C(q,t)$ $(g_C q^{-1})$ such that $\rho_C(q,t)dq$ is the amount of carbon in the interval of qualities $[q, q + dq]$ at time t $(t \geq 0)$]. Similarly, we define $\rho_b(q,t)dq$ as the amount of carbon in the decomposer biomass in the interval of qualities $[q, q + dq]$ at time t $(t \geq 0)$. Let $P(q,t)$ $(g_{biomass} t^{-1} q^{-1})$ be the decomposer growth rate on carbon of quality q, $e(q)$ (dimensionless) be the decomposer efficiency (production -to- assimilation ratio, $0 < e < 1$), and f_C $(g_C g_{biomass}^{-1})$ the carbon concentration in the decomposer biomass. P, e, and f_C have the same meaning as in Chapter 3, except that P and e now depend upon q; in principle f_C should also depend upon q but we will already here assume that f_C can be regarded as constant. The new feature that we introduce here is that carbon assimilated at some quality q' will be converted into carbon of a range of other qualities. We describe this with the function $D(q,q')$ (q^{-1}), which we call the dispersion function. There is, finally, a mortality function (or loss rate) for the decomposer biomass, $\mu_b(q)$.

As illustrated in Figure 4.1, the amount of carbon at quality q changes in two ways: there is a loss due to decomposer assimilation $f_C P(q,t)/e(q)$ and a gain due to mortality to which a quality q' contributes with a fraction $D(q,q')$ of the mortality at quality q'. Then, conservation of mass gives the following equation for the temporal change of $\rho_C(q,t)$ and $\rho_b(q,t)$

$$\frac{\partial \rho_C(q,t)}{\partial t} = -\frac{f_C P(q,t)}{e(q)} + \mu_b(q,t)\rho_b(q,t) \tag{4.1}$$

$$\frac{\partial \rho_b(q,t)}{\partial t} = \int_0^\infty f_C D(q,q')P(q',t)dq' - \mu_b(q,t)\rho_b(q,t) \tag{4.2}$$

We will now introduce the adiabatic assumption that the decomposer biomass responds rapidly relative to the changes in substrate availability; $\rho_b(q,t)$ is throughout assumed to be in a steady state or in other words $\partial\rho_b(q,t)/\partial t = 0$ (cf. 3.9a). We can then rewrite (4.1) and (4.2)

$$\frac{\partial \rho_C(q,t)}{\partial t} = -\frac{f_C P(q,t)}{e(q)} + \int_0^\infty f_C D(q,q')P(q',t)dq' \tag{4.3}$$

$$\rho_b(q,t) = \frac{1}{\mu_b(q,t)} \int_0^\infty f_C D(q,q')P(q',t)dq' \tag{4.4}$$

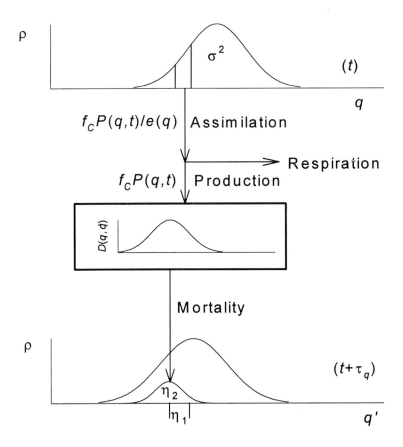

Figure 4.1 Degradation of a heterogeneous substrate. At time t the distribution of carbon $\rho_C(q,t)$ with mean value $\hat{q}(t)$ and variance $\sigma^2(t)$ is centred to the right (high quality); at a later time $t + \tau_q$ the distribution has moved to the left (low quality) with a concomitant loss of matter. The contribution of quality q to decomposer growth is $f_C P(q,t)$. Carbon from quality q is returned through mortality and distributed over the quality spectrum (q') according to $D(q',q)$. This process is asymmetric: the mean value of quality of the returned carbon has been shifted a distance η_1 ($\eta_1 > 0$) towards lower qualities and has been spread out over a range η_2 (redrawn from Bosatta & Ågren 1991a).

This equation for the substrate carbon is very general: the structure of any conventional "box-transfer model" can be reduced to a particular case of equation (4.1) or (4.3) just by choosing the appropriate kernel operator $D(q,q')$. From the definition of $D(q,q')$ as a distribution it has the following property

$$\int_0^\infty D(q,q')dq = 1 \qquad (4.5)$$

To proceed we connect the decomposer growth rate, $P(q,t)$, to the carbon density function by assuming that the utilisation of a particular fraction of carbon is proportional to its concentration (cf. (3.11)), viz.

$$P(q,t) = u(q)\rho_C(q,t) \qquad (4.6)$$

where $u(q)$ ($g_{biomass}g_C^{-1}t^{-1}$) is the decomposer growth rate per unit of carbon of quality q. The assumption (4.6) is important because it leads to linear equations. The form of equation (4.3) that we will be working with henceforth is then

$$\frac{\partial \rho_C(q,t)}{\partial t} = -\frac{f_C u(q)\rho_C(q,t)}{e(q)} + \int_0^\infty f_C D(q,q')u(q')\rho_C(q',t)dq' \qquad (4.7)$$

which has no explicit dependency on the dynamics of decomposers. Instead, we have a separate equation for the decomposer biomass (4.4). Equation (4.7) is the fundamental equation that is the basis for all our following analyses of carbon dynamics.

Problem 4.1

Consider possible ways of reformulating (4.7) to include "priming effects".

By definition of $\rho_C(q,t)$, the integral of this function over all q gives the amount of C in the substrate at time t, i.e.

$$C(t) = \int_0^\infty \rho_C(q,t)dq \qquad (4.8)$$

Using (4.7) and the property (4.5) we get

$$\frac{dC(t)}{dt} = -\int_0^\infty k(q)\rho_C(q,t)dq \qquad (4.9)$$

where

$$k(q) = f_C \frac{1 - e(q)}{e(q)} u(q) \tag{4.10}$$

is the specific decomposition rate in analogy with (3.13).

Equation (4.9) is the dynamic equation for the carbon in the substrate; it depends explicitly upon the decomposer efficiency and growth rate but not upon the dispersion function $D(q,q')$.

Since it is not possible, in general, to find a closed solution to (4.7), approximate methods must be used. There are, however, some particular choices of $D(q,q')$ that permit a complete solution to (4.7) and this is what we shall consider in the next point. A general approach is given in section 4.4.

Problem 4.2
Derive equation (4.9).

Problem 4.3
In which way does $D(q,q')$ influence the dynamics of $C(t)$?

Problem 4.4
By definition, $0 < e(q) < 1$ and $u(q) > 0$. Use this to show that $C(t)$, equation (4.8), decays monotonously in time.

Problem 4.5
Use the definitions $\rho(q,t) = \Sigma C_i(t)\delta(q - q_i)$ and $D(q,q') = \Sigma d(q,q')\delta(q - q_j)$ in order to transform (4.7) into an equation over n discrete boxes where $C_i(t)$ is the amount of carbon in box i at time t.

4.3 Two particular solutions

We will now consider two special cases. The first reason is that we need to show how simpler decomposition models can be derived as special cases. The second reason is that by judiciously choosing $D(q,q')$ we can derive exact analytical solutions. Although these choices might seem artificial, they lead to results that show important qualitative features which apply also with more complicated choices of D (cf. the use of a square-well potential in quantum mechanics).

Suppose first that

$$D(q,q') = \delta(q - q') \tag{4.11}$$

where δ is the Dirac delta function (see Appendix 1). Equation (4.11) implies that the decomposers do not change the quality of the assimilated carbon, i.e.

assimilated carbon of quality q is returned with quality q. Since there is no interaction between qualities (4.11) should give us back the equations for homogeneous substrates of Chapter 3. Inserting (4.11) into (4.7) and using the property (A1.5) we get

$$\frac{\partial \rho_C(q,t)}{\partial t} = -k(q)\rho_C(q,t) \tag{4.12}$$

which has the solution

$$\rho_C(q,t) = \rho_C(q,0)e^{-k(q)t} \tag{4.13}$$

and so, each "box" q decays exponentially in time with its own specific decay rate $k(q)$ defined by (4.10).[4]

Problem 4.6

A sum of exponential functions has sometimes been proposed (e.g. Minderman, 1968) as a model to represent the dynamics of carbon in decomposing litter structures in which each exponential term represents the dynamics of a different kind of carbon (e.g. sugars, lignin, etc.). Show that starting with equation (4.13) and defining $\rho(q,0) = \Sigma C_i \delta(q - q_i)$ you can derive this kind of behaviour.

A more difficult problem is created if we choose

$$D(q,q') = \delta(q-q') + \eta_1(q)\frac{\partial}{\partial q}\delta(q-q') \tag{4.14}$$

where $\eta_1(q)$ is a function describing the rate of change in dispersion. Equation (4.14) states that for each cycle of carbon through the decomposers the quality is shifted a distance η_1 towards lower qualities.

With (4.14) and using (A1.5), (4.7) transforms into[5]

$$\frac{\partial \rho_C(q,t)}{\partial t} = -\frac{1-e(q)}{e(q)}f_C u(q)\rho_C(q,t) + f_C\frac{\partial}{\partial q}\left[\eta_1(q)u(q)\rho_C(q,t)\right] \tag{4.15}$$

The first term of (4.15) represents catabolic losses of carbon, while the second can be visualised as a net rate of flow of carbon towards lower qualities (humification).

If, initially, all carbon in the substrate is of quality q_0, i.e. if $\rho_C(q,0) = C_0\delta(q - q_0)$ where C_0 is the initial amount of carbon, the solution to (4.15) is

$$\rho_C(q,t) = C_0 e^{-\int_q^{q_0} \frac{1-e(q')}{\eta_1(q')e(q')}dq'} \delta(q_t - q) \tag{4.16}$$

where $q_t = q(t)$ is the solution to the following equation

$$\frac{dq_t}{dt} = -f_C \eta_1(q_t)u(q_t) \tag{4.17}$$

Thus, with η_2 (cf. Figure 4.1) and all higher order moments of D (see (4.24) below) equal to zero, the initial δ-distribution remains a δ-distribution at all times: only the mean value of quality changes in time and this according to equation (4.17). This equation simply expresses that quality changes with the rate at which the decomposers can use the substrate; the rate of degradation in quality is also proportional to the mean change in quality η_1. The quantities in the right hand side of (4.17) are all positive, implying that quality will decrease monotonously from its initial value q_0 until a final value where decomposer growth rate $u(q) = 0$.

Integrating (4.16) over q we get the amount of carbon in the substrate at time t

$$C(t) = C_0 e^{-\int_{q_t}^{q_0} \frac{1-e(q)}{\eta_1(q)e(q)}dq} \tag{4.18}$$

Inserting (4.16) into (4.9), integrating and using (4.18) we get

$$\frac{dC(t)}{dt} = -k(q_t)C(t) \tag{4.19}$$

and so, $k(q_t) = k(t)$ becomes the specific decomposition rate of the substrate which is no longer a constant as in Chapter 3 but a function of time. However, the important result here is that we have an alternative way of studying the carbon dynamics - instead of looking at changes in C with time we look at changes in C as a function q. We started with one equation (4.7) in two variables, q and t, and ended up with two independent equations (4.17) and (4.18), each in only one variable. These equations are, of course, much simpler to analyse. However, it is more important that this result suggests that even in other situations it should be profitable to look at changes in quality over time and changes in carbon amounts as a function of quality as two independent processes. It is also important to recognise that the two equations also have separated the influence of the decomposer properties; $C(q)$ depends on $e(q)$ and $\eta_1(q)$ and $q(t)$ depends on $u(q)$ and $\eta_1(q)$. Thus, we do not need to know $e(q)$ if we want to know how quality is changing in time and we do not need to know $u(q)$ if we want to know how carbon is changing with quality. This last observation is significant because if we want to introduce climatic factors explicitly, it is foremost the decomposer growth rate, i.e. $u(q)$, that should contain these factors. This suggests that decomposition descriptions based on quality changes should be more fundamental than those based on a temporal dimension as trivial influences of climatic variations are removed.

Problem 4.7

Verify that (4.16) is a solution to (4.15).

Problem 4.8

Suppose that $e(q) = e_0$, $u(q) = u_0q$, and $\eta_1(q) = \eta_{10}$, where e_0 and u_0 are two positive constants. Use (4.17) and (4.18) to calculate $q(t)$, $C(t)$ and $k(t)$. What are the values of these variables when $t = \infty$?

Problem 4.9

Suppose that $D(q,q') = \delta(q - q') - \eta_1\delta'(q - q') + \eta_2\delta''(q - q')$ where η_1 and η_2 are constants. Derive the dynamic equation for $\rho(q,t)$.

4.4   The moment expansion

It is quite clear that finding a general solution for (4.7) for arbitrary decomposer functions is a hopeless task. It is also clear that more complicated functions for D than those used in section 4.3 most likely will lead to equations that are extremely difficult, if at all possible, to solve analytically. What we would like to have is a method that allows us to systematically approximate (4.7). One such method is to expand functions in some type of series. We can do this by operating with the moments of these functions. The equations for the moments are simpler and the moments could be easier to observe; several of them have direct, simple biological interpretations. We therefore define the following functions:

The total amount of carbon in the substrate, $C(t)$ (see also (4.8))

$$C(t) = \int_0^\infty \rho_C(q,t)dq \tag{4.20}$$

The average quality of the substrate, $\hat{q}(t) = \hat{q}_t$

$$\hat{q}(t) = \frac{1}{C(t)} \int_0^\infty q\rho_C(q,t)dq \qquad (4.21)$$

The variance in substrate quality, $\sigma^2(t)$

$$\sigma^2(t) = \frac{1}{C(t)} \int_0^\infty [q - \hat{q}(t)]^2 \rho_C(q,t)dq \qquad (4.22)$$

Higher order moments, $\mu_n(t)$, for which there are no simple biological interpretations

$$\mu_n(t) = \frac{1}{C(t)} \int_0^\infty [q - \hat{q}(t)]^n \rho_C(q,t) \qquad (4.23)$$

Functions $\eta_n(q)$, which define moments of the distribution of the carbon that passes back from the decomposers to the substrate

$$\eta_n(q) = \int_0^\infty (q - q')^n D(q',q)dq' \qquad (4.24)$$

In particular, the zeroth order moment is the total amount of carbon returned, so by definition, cf. equation (4.5)

$$\eta_0 = \int_0^\infty D(q',q)dq' = 1 \qquad (4.25)$$

The first order moment, $\eta_1(q)$, gives the average change in quality of carbon assimilated at quality q when that carbon is returned to the substrate. Note, that with the definition used, η_1 is positive when the substrate quality is decreasing during decomposition. Similarly, $\eta_2(q)$ is the variance in the quality of the carbon returned to the substrate.

The dynamic equations for $C(t)$, $\hat{q}_t$, $\sigma^2(t)$, and higher order moments can be derived in a straight-forward manner by taking the derivatives with respect to time of equations (4.20)-(4.23) and then replacing $d\rho_C/dt$ with equation (4.7). By expanding the appropriate functions around $\hat{q}_t$ we derive a system of ordinary differential equations in the moments (see Appendix 2) . This system can be solved if we truncate the series at some convenient point. Our knowledge of the system and the possible importance of higher order moments determine where such a truncation should be done. When truncating at the second order moment, we get (with $k(q)$ from (4.10))

$$\frac{dC(t)}{dt} = -\left[k(\hat{q}) + \frac{\sigma^2(t)}{2}\frac{\partial^2 k(\hat{q})}{\partial \hat{q}^2}\right]C(t) \tag{4.26}$$

$$\frac{d\hat{q}(t)}{dt} = -f_C\eta_1(\hat{q})u(\hat{q}) - \left[\frac{\partial k(\hat{q})}{\partial \hat{q}} + \frac{1}{2}f_C\frac{\partial^2(\eta_1 u)}{\partial q^2}\right]\sigma^2(t) \tag{4.27}$$

$$\frac{d\sigma^2}{dt} = f_C\eta_2(\hat{q})u(\hat{q}) - \left[2f_C\frac{\partial(\eta_1 u)}{\partial \hat{q}} - \frac{1}{2}f_C\frac{\partial^2(\eta_2 u)}{\partial \hat{q}^2}\right]\sigma^2(t) \tag{4.28}$$

The moment expansion displays some qualitative effects of dispersion. When $\eta_n = 0$ for all $n \geq 2$, a carbon distribution starting with zero variance will remain with zero variance forever and (4.27) reduces to (4.17). If $\eta_2 > 0$, then even if the variance is zero initially, the first term in (4.28) ensures that $\sigma^2 > 0$ for $t > 0$; the increase in σ^2 will probably level off towards some quasi-stationary state as terms of the opposite sign become more dominant in equation (4.28). Simultaneously, the average quality, $\hat{q}_t$ is likely to decrease more rapidly; in equation (4.27) the term $dk/d\hat{q}_t$ should be positive whereas it is difficult to forecast the sign of $d^2(u\eta_1)/d\hat{q}_t^2$. The implications for the substrate carbon are more complicated as the effects of the higher order moments are coupled to second and higher order derivatives of k for which the sign is less easily predictable; for example, with the models for $e(q)$ and $u(q)$ used in Appendix 2 (where we will study numerical solutions to the moment expansion), $d^2k/d\hat{q}_t^2$ increases from negative values at high q to positive values at low ones, implying that for substrates of low quality dispersion accelerates carbon loss but for high quality substrates carbon losses are retarded.

In the coming practical applications we shall retain only the first order terms in our equations, i.e. (4.26) and (4.27) with $\sigma^2 = 0$. Within this approximation the solution to (4.26) is given by

$$C(t) = C_0 e^{-\int_0^t \frac{1-e(\hat{q}_\tau)}{e(\hat{q}_\tau)}f_C u(\hat{q}_\tau)d\tau} = C_0 e^{-\int_{\hat{q}_t}^{q_0}\frac{1-e(q)}{\eta_1(q)e(q)}dq} \tag{4.29}$$

Note that (4.29) is a general solution (within the first order approximation) without any assumptions about specific models for e, u and D giving us back equations derived previously with special models for D; (4.27) equals (4.17) and (4.29) equals (4.18).

Our formalisation of carbon dynamics is thereby complete; we now need to define the dynamics of N, P and S.

Problem 4.10

Convince yourself that η_1 of (4.24) is positive if quality is decreasing during decomposition.

Problem 4.11

Derive equation (4.27).

4.5 Basic equations for N, P and S

We define a density function of element n (= N, P, or S), $\rho_n(q,t)$ $(g_n q^{-1})$, such that $\rho_n(q,t)dq$ is the amount of n in the interval of qualities $[q, q + dq]$ at time t. This definition is identical to the one for carbon, but the dynamic equation governing it will be different. We assume that the decomposer community is carbon limited, such that the turnover of n is regulated by carbon utilisation which means that whenever a given amount of carbon is taken by a decomposer from the quality q, an amount of n corresponding to the concentration ρ_n/ρ_C, is removed at the same quality. Dead decomposers return element n to the substrate with their n concentration, $f_n(q)$ and, as for carbon, with no delay. Since different organisms are likely to decompose substrates of different qualities, the n concentration of the dead decomposers can vary with quality. The dispersion function will be the same for both carbon and nitrogen. Thus, the dynamic equation for ρ_n is

$$\frac{\partial \rho_n(q,t)}{\partial t} = -\frac{f_C u(q)}{e(q)} \rho_n(q,t) + \int_0^\infty f_n(q)D(q,q')u(q')\rho_C(q',t)dq' \quad (4.30)$$

By definition, the total amount of n in the substrate at time t is

$$n(t) = \int_0^\infty \rho_n(q,t)dq \quad (4.31)$$

Integrating (4.30) over q and using (4.5) we get

$$\frac{dn(t)}{dt} = -\int_0^\infty f_C \frac{u(q)}{e(q)} P_n(q,t) + \int_0^\infty f_n(q)u(q)\rho_C(q,t)dq \qquad (4.32)$$

and expanding the functions around $\hat{q}_t$ and using (4.20) and (4.31) gives

$$\frac{dn(t)}{dt} = -\frac{f_C u(\hat{q})}{e(\hat{q})} n(t) + f_n(\hat{q})u(\hat{q})C(t) \qquad (4.33)$$

which should be compared with the equation for the homogeneous substrate (3.12b). If at the beginning of decomposition, the second term in the right hand side of (4.33) predominates over the first, n will accumulate in the substrate until a critical value of quality, q_c, is reached; at that moment, $dn/dt = 0$, the n:C ratio takes the critical value $r_{nc} = e(q_c)f_n(q_c)/f_C$ and $n(t)$ reaches a maximum value n_c; after that, a net release of n is started from the substrate.

When the initial nutrient concentration is r_{0n}, the solution to (4.33) is (f_n has been assumed constant)

$$n(t) = \left[\frac{f_n}{f_C} - \left(\frac{f_n}{f_C} - r_{0n} \right) e^{-\int_0^t f_C u(\hat{q}_\tau)d\tau} \right] C(t) \qquad (4.34)$$

where $C(t)$ is given in (4.29). The expression in square brackets in (4.34) is the n:C ratio, $r_n(t) = n(t)/C(t)$ of the substrate. The integral in (4.34) is proportional to the amount of decomposer biomass formed during the time interval $[0,t]$ and, the larger this amount, viz., the larger the amount of substrate assimilated, the more $r_n(t)$ approaches f_n/f_C. The integral in (4.34) is therefore a monotonically increasing function of time and as a consequence (and without any assumptions about specific models for e, u, and D) it follows that r_n is also monotonically increasing with time and bounded in the interval $r_{0n} < r(t) < f_n/f_C$.[6] There is the alternative of expressing (4.34) in terms of quality using (4.17), $f_C u(\hat{q}_t)d\tau$ can be substituted with $d\hat{q}_t/\eta_1$

$$r_n(t) = \frac{f_n}{f_C} - \left(\frac{f_n}{f_C} - r_{0n} \right) e^{-\int_{\hat{q}_t}^{q_0} \frac{1}{\eta_1(q)}dq} \qquad (4.35)$$

which shows that the final value of r_n, $r_n(\infty)$ can depend not only upon decomposer properties, as was the case for the homogeneous substrate, but also upon the initial quality of the substrate. It is noteworthy that the only decomposer property that determines $r_n(\infty)$, besides f_n/f_C, is $\eta_1(q)$. It is clear that $u(q)$ can play no role as it only controls the rate at which the system approaches some state. The fluxes of carbon and nitrogen from the substrate to the decomposer biomass or the environment are equally affected by $e(q)$. Hence, $e(q)$ cannot change the N:C ratio of the substrate. Since $\eta_1(q)$ is the average displacement in quality for each cycle through the decomposer biomass, $dq/\eta_1(q)$ is the fraction of a cycle undergone when quality is changing with dq (it is easy to understand this when η_1 is constant). The integral in (4.35) is, therefore, the number of cycles a substrate has passed through the decomposer biomass when decreasing its quality from the initial value q_0 to the current $\hat{q}_t$. And the more cycles a substrate has undergone, the more its properties will approach those of the decomposer biomass.

We now have the basic elements to describe the dynamics of C, N, P and S.

Problem 4.12
Verify that (4.34) satisfies (4.33).

Problem 4.13
Using the decomposer models of Problem 4.15, use (4.33) and (4.35) to calculate q_c, t_c, and r_{nc} (assume f_n constant).

4.6   One litter cohort

It is convenient, as was done in Chapter 3, to define $g(t)$ and $h_n(t)$ as the amounts of carbon and nutrient relative to the initial amount of carbon in the litter cohort. We obtain the dynamics of $g(t)$ and $h_n(t)$ from (4.26) and (4.33) (remember that we have chosen $\eta_2 = \sigma_2 = 0$)

$$\frac{dg(t)}{dt} = -\frac{1-e(\hat{q})}{e(\hat{q})} f_C u(\hat{q})g(t) = -k(\hat{q})g(t) \tag{4.36}$$

$$\frac{dh_n(t)}{dt} = -\frac{f_C u(\hat{q})}{e(\hat{q})} h_n(t) + f_n u(\hat{q})g(t) \tag{4.37}$$

where, for simplicity, f_n is assumed to be constant. In (4.36) $f_C ug/e$ is the gross consumption rate (assimilation) of C from the litter cohort while $f_C ug$ goes to the production of new biomass; the difference $(1/e - 1)f_C ug$ is the rate of carbon losses from the litter cohort (respiration). In a similar way the first term of the right hand side of (4.37) corresponds to the rate of gross minerali-

sation of element n from the substrate while the second is the rate of incorporation of n into the decomposer biomass (see section 7.1); the difference gives
the rate of net mineralisation of element n from the litter cohort.

Dividing (4.36) and (4.37) by (4.27) we get

$$\frac{dg(q)}{dq} = \frac{1-e(q)}{\eta_1(q)e(q)}g(q) \tag{4.38}$$

$$\frac{dh_n}{dq} = \frac{1}{\eta_1(q)e(q)}h_n(q) - \frac{f_n}{f_C\eta_1(q)}g(q) \tag{4.39}$$

which is a convenient alternative way of writing the dynamics of g and h_n in
terms of q instead of t (and where, for simplicity, we just write q instead of
$\hat{q}_t$). The use of q instead of t (the q-representation) is convenient for several
reasons: i) the equations are mathematically simpler, ii) the growth function
$u(q)$ disappears from the equations and with it all externally related factors
such as climate, and iii) the properties of the long-term dynamics become
easier to analyse (this is achieved by studying the behaviour of certain functions in the limit of $q = 0$). The q-representation provides, thus, a more
intrinsic description of the decomposition process.

Problem 4.14
The q-representation works best with the first order approximation in the moment
expansion. Why?

Equations (4.36), (4.37) and (4.27) or, alternatively (4.27), (4.38) and
(4.39) are basically all that we shall use in the forthcoming analyses. The
assumptions made to derive these equations are: i) decomposers are C-
limited, ii) no delays between assimilation and mortality, iii) no dispersion in
$D(q,q')$, i.e. η_2 and all higher order moments are zero. To use these equations
we assume then that: i) the decomposers have available inorganic sources of n
during the immobilisation phase, ii) for medium and long-term dynamics the
properties of the kernel operator $D(q,q')$ dominate in (4.3) and lag effects can
be neglected, i.e. we restrict the attention to time scales where short-term
variations in decomposer biomasses can be neglected (it is not necessary to
represent the decomposer biomass explicitly but it will follow the substrate
carbon level, (4.4)), iii) to define properties of the organic matter, dispersion
effects in individual litter cohorts are of less importance than dispersion

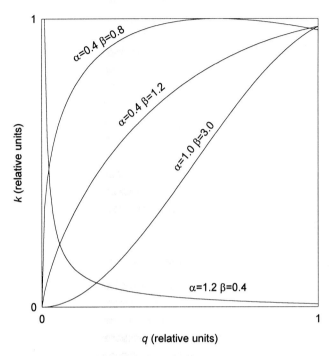

Figure 4.2 Some curves of possible relationships between specific decomposition rate, k, and substrate quality, q. All curves are derived with $e(q) = e_1 q^\alpha$, and $u(q) = u_0 q^\beta$, and scaled such that $k_{max} = 1$ (from Bosatta & Ågren 1985).

effects produced by the mixture of cohorts (note that the idea of "cohort" is useful as long as the quality of the incoming litter is well defined, i.e. a δ or δ -like distribution centred around q_0 (there is here an interesting analogy to the "particle" concept in physics), and iv) that for each cycle of carbon through the decomposers the quality is shifted on the average towards lower qualities.

Equations (4.36) and (4.37) and the general solution to them (4.18) and (4.35), can be interpreted as an infinite set of models depending upon three general decomposer functions, viz. $e(q)$, the decomposer efficiency in substrate utilisation, $u(q)$, the decomposer growth rate per unit of carbon, and $\eta_1(q)$ the displacement in quality. Details of the decomposition process require that these functions be specified. There are, however, qualitatively important results that can be obtained only from the general form of $e(q)$, $u(q)$, and $\eta_1(q)$. Decomposers degrade a material both in quantity and in quality; if the degradation in quality, dq/dt, is more rapid than in quantity, dg/dq, then we get a situation where the substrate becomes undegradable but a finite amount remains, i.e. $g(q)$ remains finite when the decomposition has proceeded to the situation where the substrate quality is zero. This degradation in both quantity and quality is also reflected in the specific decomposition rate, $k(q)$, which may be both increasing or decreasing with q, Figure 4.2, where we for the moment allow other decomposer functions than those defined in 4.41, see below. In order that all the substrate eventually becomes decomposed, $g(q = 0) = 0$, it is therefore necessary that the degradation in quantity is more rapid than in quality. In other words, since the degradation in quantity is inversely proportional to $\eta_1(q)e(q)$ (equation 4.38), it is necessary that $\eta_1(q)e(q)$ goes to zero sufficiently rapidly at $q = 0$, more precisely at the q at which $\eta_1(q)e(q)$ goes to zero; for convenience we let this happen at $q = 0$. This means that the long term behaviour of g is controlled by the behaviour of the decomposer efficiency $\eta_1(q)e(q)$, in the limit of $q = 0$, see (4.29). Another general result, which depends upon the behaviour of the function $\eta_1(q)u(q)$ at the point where $\eta_1(q)u(q) = 0$, follows by formal integration of equation (4.27) (with $\sigma^2 = 0$)

$$t = f_C \int_q^{q_0} \frac{dq'}{\eta_1(q')u(q')} \tag{4.40}$$

If $\eta_1(q)u(q)$ is such that the integral (4.40) is finite for all $q \geq 0$, then there exists a $t = t_{max}$ such that, for $t > t_{max}$ there is no solution to (4.40). This means that the quality, q, will go to zero in a finite time. Hence, there are two qualitative distinct possibilities - the decomposition process can proceed for an infinitely long time or it can stop after a finite time. The function $u(q)$ is also responsible for the decrease in the specific decay rate with time or, alternatively, with the decrease in the substrate quality; (4.36) shows that the decay rate in principle decreases with decreasing decomposer growth rate or with increasing decomposer efficiency. However, since both decomposer efficiency and decomposer growth rate decrease with decreasing substrate quality we have the interesting possibility that the decay rate can increase with decreasing substrate quality. This occurs if the efficiency function, $e(q)$, decreases more rapidly than the growth rate function, which means that although the decomposers with decreasing substrate quality are increasingly slow in their use of the substrate, they burn away even more of the substrate as a result of the decrease in efficiency, Figure 4.2.

Problem 4.15
Using the definitions $e(q) = e_1 q^\alpha$, $u(q) = u_0 q^\beta$ (α, $\beta > 0$), and $\eta_1(q) = \eta_{10}$, calculate $g(q)$ and $q(t)$. Analyse the long-term behaviour of these functions for different values of α and β.

We will now continue the analysis by using some specific decomposer models.

4.7 Models for decomposer functions

We now need to look for models for the three decomposer functions, $e(q)$, $u(q)$, and $D(q,q')$. There is only limited experimental evidence available to guide us. However, we can argue on rather general grounds about possible models. In addition, we will chose models that simplify our calculations.

Let us start with decomposer efficiency, e. Degradation of a complex molecule requires more enzymes than degradation of a simple one. The costs to use a complex molecule should, therefore, be higher and the decomposer efficiency should be an increasing function of substrate quality. The simplest increasing function is a linear function, hence our choice of model will be

$$e(q) = e_0 + e_1 q \tag{4.41}$$

Additional simplifications obtain when either of e_0 or e_1 is 0.

The decomposer growth rate, $u(q)$, is by definition a growing function of quality. In the limit of physically perfectly free access to a carbon atom within a substrate, the only meaningful way of defining quality requires that u is a linear function of q. However, the carbon atoms are physically protected from decomposition. Such protection comes from, on one hand, the three-dimensional structure of the molecule where the particular carbon atom resides and, on the other hand, from interactions with the surroundings, e.g. adsorption to or inclusion in the soil matrix. At the moment we do not know how these interactions vary with quality. The more resistant substrates will, however, have a longer residence time in the soil and therefore have a higher probability to be fixed in the soil matrix. The effect should be that in relative terms the growth rate on low quality substrates is lower than expected just from its chemistry. We have chosen to represent this with a power function in q,

$$u(q) = u_0 q^\beta \tag{4.42}$$

The parameter u_0 is a basic growth rate. This is also the parameter that will change in response to e.g. temperature and water availability. The other parameter, β, defines the strength of the physical protection and is ≥ 1 with the important limiting case of 1 for a perfectly free substrate.

We will for the moment not try to define any model for $D(q,q')$ but be content to look at models of its moment, even restricting ourselves to only the first moment, η_1. As decomposition proceeds, the transformations of the substrate will increasingly be between complex heterogeneous substrates, and thus it is reasonable to assume that these will become increasingly alike. Hence, η_1 that describes the changes in quality, should be an increasing function of quality. We also require that $D(q,q') = 0$ for $q \leq 0$ to prevent negative qualities being created. For simplicity, we choose the following function

$$\eta_1(q) = \eta_{10} + \eta_{11}q^\alpha \qquad (4.43)$$

Again, we will pay specific attention to the two special cases where either of η_{10} or η_{11} is 0.

4.8   Model I

In this section, we will use what we will call the Model I for decomposer functions. This is the model first analysed by us. It is characterised by $\eta_1(q) = \eta_{10}$ and $e(q) = e_0 + e_1 q$.

4.8.1   *One litter cohort*

Inserting (4.41) and (4.42) in (4.36) and integrating gives

$$g(q) = e^{\frac{q_0-q}{\eta_{10}}} \left[\frac{e_0 + e_1 q}{e_0 + e_1 q_0} \right]^{\frac{1}{\eta_{10}e_1}} \qquad (4.44)$$

and

$$h_n(q) = r_n(q)g(q) = \left[\frac{f_n}{f_C} + \left(r_{0n} - \frac{f_n}{f_C} \right) e^{\frac{q-q_0}{\eta_{10}}} \right] \left[\frac{e_0 + e_1 q}{e_0 + e_1 q_0} \right]^{\frac{1}{\eta_{10}e_1}} e^{\frac{q_0-q}{\eta_{10}}} \quad (4.45)$$

Figure 4.3 shows the dynamic behaviour of g and h_n; the parameters used in the figure are typical of C and N in Scots pine needles (see Chapter 5).

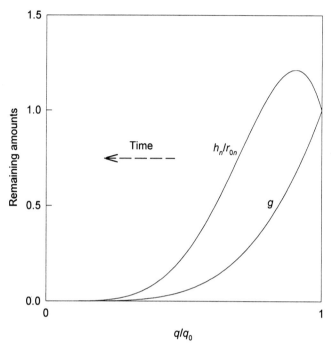

Figure 4.3 Behaviour of g and h_n/r_{0n} (remaining amounts of C and n relative to initial amounts of C and n) as a function of quality for a litter cohort. The model used for decomposer efficiency is $e(q) = e_1q$ and $\eta_1(q) = \eta_{10}$. $g(q)$ decays monotonously to zero $g(q = 0) = 0$. $h_n(q)$ increases first (immobilisation) until a critical value of quality, q_c, is reached in the litter cohort; afterwards, a net release of n is started. Parameter values are: $q_0/\eta_{10} = 1$, $e_1\eta_{10} = 0.2$, $(f_n/f_C)/r_{0n} = 10$.

Problem 4.16
Analyse the effects of different parameter values upon $g(q)$ and $h_n(q)$ of Figure 4.3.

In applications, a practical way of eliminating the time scale is by representing the amount of n in terms of the amount of carbon or, alternatively, in terms of mass losses of the decomposing material. This can be achieved by elimination of the variable q between equations (4.44) and (4.45), obtaining in this way h_n as a function of g or alternatively, as a function of $1 - g$ (mass loss).

The solution to (4.27) with definition (4.42) for $u(q)$ is

$$q(t) = \frac{q_0}{\left[1 + (\beta - 1)\eta_{10} f_C q_0^{\beta - 1} t\right]^{\frac{1}{\beta - 1}}} \tag{4.46}$$

which shows that quality goes to zero as $t^{-1/(\beta - 1)}$. Equation (4.46) provides the link between the t and q scales; inserting it into (4.44) and (4.45) we get the evolution of g and h_n over time.

We now have developed basic equations to describe the dynamics of C and nutrient in single litter cohorts for one set of models for the decomposer functions. We can now proceed to the formulation for soil organic matter.

4.8.2   *Several litter cohorts*

Let $I(t)$ denote the rate of input of any litter fraction (plant component) expressed in units of carbon ($g_C cm^{-2} t^{-1}$), and let a denote the age of a litter cohort. Since the first cohort entered the soil at $t = 0$, the age of the oldest cohort is t and $I(t - a)da$ is the initial amount of carbon in the litter cohort of age a. Using the definitions of g and h_n, we calculate the amounts of carbon and nutrients in the organic matter formed by the litter fraction by summing over all cohorts:

$$C(t) = \int_0^t I(t - a) g(q_a) da \tag{4.47}$$

$$n(t) = \int_0^t I(t - a) h_n(q_a) da \tag{4.48}$$

Sometimes it is convenient to integrate over q instead of over ages. When a is substituted by q we have to substitute da in (4.47) and (4.48) with $-dq/[\eta_{10} f_C u(q)]$ (4.27) and the limits of integration 0 and t with q_0 and q_t, respectively, where q_0 is the quality of the youngest and q_t of the oldest cohort, respectively.

Assuming that the rate of litter input is constant, we get

$$C(t) = \frac{I_0}{\eta_{10} f_C u_0} \int_{q_t}^{q_0} \frac{1}{q^{\beta}} \left[\frac{e_0 + e_1 q}{e_0 + e_1 q_0} \right]^{\frac{1}{\eta_{10} e_1}} e^{\frac{q_0 - q}{\eta_{10}}} dq \tag{4.49}$$

The integrand in (4.49) can be interpreted as the distribution of carbon over quality. In other words, the integrand tells us the relative amounts of carbon of different qualities that we should expect in the organic matter.[7]

The properties of the integrand in (4.49) vary greatly depending upon parameter values, Figure 4.4. A first distinction should be made between $e_0 > 0$ and $e_0 = 0$. When $e_0 > 0$, the integrand always diverges for small q which reflects the fact that all cohorts are contributing with a finite remainder. For larger values of q, the integrand can (i) decrease monotonically, (ii) have one minimum, or (iii) have one minimum and one maximum. Similarly, there exist three different situations for $e_0 = 0$: (i) when $\beta \eta_{10} e_1 > 1$ the integrand diverges for small q; (ii) the integrand goes to zero over a maximum with decreasing q; (iii) the integrand decreases monotonically to zero with q. The behaviour of the integrand for small q is of particular interest, because a divergence of the integrand implies that with the ageing of the soil an increasing amount of the carbon will be tied up in the most recalcitrant substrates. Our formalism makes the transformation of the soil organic matter continuous and without a classification into, e.g. litter and humus. If, for other reasons than morphological, it is of interest to differentiate the substrate into categories according to its degree of decomposition, Figure 4.4 shows possibilities for such a classification. For some combinations of parameters, the distribution of carbon over qualities has a minimum and/or a maximum. These extremes could be used as dividing points for the classification. However, when the distribution is changing monotonically no such criteria for a classification exist.

Explicit calculation of the integrals (4.47) and (4.48) provides information about the dynamics of C and nutrients in the organic matter resulting from one type of litter. In order to calculate the dynamics in the whole soil, we might need to add the contributions of (4.47) and (4.48) from different litter fractions.

4.9   Model II

In part the analysis in section 4.8 was based on the earliest parameterisations of the decomposer functions and was a consequence of the way we first derived the basic equations. There are some problems with that parameterisation. First of all, choosing $\beta = 1$ (a substrate without physical obstruction to the decomposers) leads to a specific decomposition rate that increases with time for several decomposer efficiency models. This is biologically less attractive. Secondly, by choosing $\eta_1(q)$ = constant, we assume that $D(0,q') \neq 0$, i.e. finite amounts of carbon are added to an undecomposable quality. This may be less desirable. The current derivation

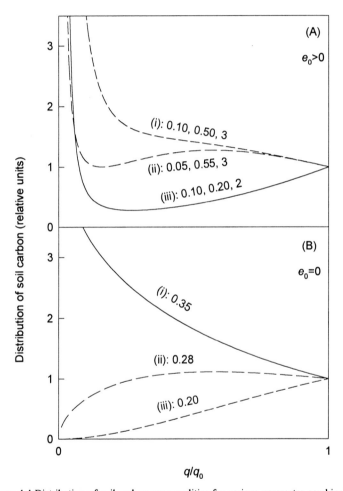

Figure 4.4 Distribution of soil carbon over qualities for various parameter combinations. The figures are scaled relative to initial quality, such that the distribution is 1 for $q/q_0 = 1$. (i), (ii), and (iii) refer to corresponding cases in the text. (A) $e_0 > 0$. $\beta = 3$. The values on the lines are the parameter combinations e_0, $e_1 q_0$, and q_0/η_{10}, respectively. (B) $e_0 = 0$. $\beta = 3$. The values on the lines are the parameter $e_1 \eta_{10}$. (from Ågren & Bosatta 1987).

of (4.7) and analysis through a moment expansion makes it possible to avoid these two problems. By choosing $\eta_{11} > 0$ and $\eta_{10} = 0$, the second problem is avoided. As it turns out, we also get rid of the first one. However, in order to solve some of our equations we need to introduce the simplification $e_1 = 0$; i.e. the decomposer efficiency is independent of substrate quality. Although this is certainly not strictly the case, it is valid when decomposer efficiency only varies between narrow limits. The value of α in (4.43) now comes to play a critical role. When $\alpha < 1$, the integration of (4.38) leads to an unde-composable residue and when $\alpha \geq 1$ we get a complete decomposition. We can understand this behaviour in terms of D. If D approaches 0 sufficiently steeply when $q \to 0$, more carbon of low quality will be packed into a region where its decomposition rate is slow. The precise condition on the steepness of D gives $\alpha = 1$ as the border between the two types of behaviour. For simplicity, we will use $\alpha = 1$ in the rest of our analyses. Thus, Model II is characterised by $\eta_1(q) = \eta_{11}q$ and $e(q) = e_0 + e_1 q$, but with $e_1 = 0$ in most applications.[8]

Problem 4.17
Show how the behaviour of $D(q,q')$ in the limit of $q \to 0$ determines whether an undecomposable residue will be created or not.

4.9.1   *One litter cohort*

We can now solve (4.41) to get the development of carbon

$$g(q) = \left[\frac{\eta_{10} + \eta_{11}q}{\eta_{10} + \eta_{11}q_0} \right]^{\frac{1-e_0}{\eta_{11}e_0}} \to \left[\frac{q}{q_0} \right]^{\frac{1-e_0}{\eta_{11}e_0}} \text{ as } \eta_{10} \to 0 \qquad (4.50)$$

The nutrient:carbon ratio follows from integration of (4.35)

$$r(q) = \frac{f_n}{f_C} - \left(\frac{f_n}{f_C} - r_0 \right) \left[\frac{\eta_{10} + \eta_{11}q}{\eta_{10} + \eta_{11}q_0} \right]^{\frac{1}{\eta_{11}}} \to \frac{f_n}{f_C} - \left(\frac{f_n}{f_C} - r_0 \right) \left[\frac{q}{q_0} \right]^{\frac{1}{\eta_{11}}} \text{ as } \eta_{10} \to 0$$

$$(4.51)$$

and the development of quality from integration of (4.27), cf. (4.46)

$$q(t) = \frac{q_0}{\left[1 + \beta\eta_{11}f_C u_0 q_0^\beta t \right]^{\frac{1}{\beta}}} \qquad (4.52)$$

We see that, in contrast to the previous section, the choice of $\eta_{10} = 0$ always leads to complete decomposition of a single litter cohort and that its nutrient:carbon ratio converges towards that of the decomposers. When we add several (infinitely many) cohorts together, this may change.

4.9.2   *Several litter cohorts*

Assuming again that the litter input rate is constant, we can evaluate (4.47) and (4.48).

$$
C(t) = \frac{I_0}{f_C \eta_{11} u_0} \int_{q_t}^{q_0} \left[\frac{q}{q_0} \right]^{\frac{1-e_0}{\eta_{11}e_0}} \frac{dq}{q^{\beta+1}} = \frac{I_0}{f_C \eta_{11} u_0 q_0{}^{\beta}} \frac{1}{\frac{1-e_0}{\eta_{11}e_0} - \beta} \left[1 - \left(\frac{q_t}{q_0} \right)^{\frac{1-e_0}{\eta_{11}e_0} - \beta} \right]
$$

$$(4.53)$$

and

$$
n(t) = \frac{I_0}{f_C \eta_{11} u_0 q_0{}^{\beta}} \left[\frac{f_n}{f_C} \frac{1}{\frac{1-e_0}{\eta_{11}e_0} - \beta} \left[1 - \left(\frac{q_t}{q_0} \right)^{\frac{1-e_0}{\eta_{11}e_0} - \beta} \right] \right.
$$

$$
\left. - \left(\frac{f_n}{f_C} - r_0 \right) \frac{1}{\frac{1}{\eta_{11}e_0} - \beta} \left[1 - \left(\frac{q_t}{q_0} \right)^{\frac{1}{\eta_{11}e_0} - \beta} \right] \right]
$$

$$(4.54)$$

With these models, we can calculate exactly the carbon and nutrient amounts for the sum of the cohorts. We can also see that if β is sufficiently small, i.e. the physical protection is not too large,

$$
\beta < \frac{1-e_0}{\eta_{11}e_0}
$$

$$(4.55)$$

only finite amounts of carbon and nutrients will accumulate, whereas when (4.55) does not hold, infinite amounts will accumulate with time (the term in q_t/q_0 in (4.53) and (4.54) diverges as $q_t \to 0$). The importance of physical protection for the accumulation is clear if we interpret the deviation of β from 1 as a measure of such a protection.

We also get a fairly simple expression for the N:C ratio of the soil

$$\frac{n_L}{C_L} = \frac{f_n}{f_C} - \left(\frac{f_n}{f_C} - r_{n0}\right)\left(1 - \frac{e_0}{1 - \beta\eta_{10}e_0}\right) \frac{1 - \left(\frac{q_t}{q_0}\right)^{\frac{1}{\eta_{11}e_0} - \beta}}{1 - \left(\frac{q_t}{q_0}\right)^{\frac{1-e_0}{\eta_{11}e_0} - \beta}} \quad (4.56)$$

We can also obtain a closed expression for the mean quality of the soil organic matter

$$\bar{q}(t) = \frac{\int_0^t q_a g(q_a)\,da}{\int_0^t g(q_a)\,da} = \frac{1 - e_0 - \eta_{11}e_0\beta}{1 - e_0 - \eta_{11}e_0(\beta - 1)} \frac{1 - \left(\frac{q_t}{q_0}\right)^{\frac{1-e_0}{\eta_{11}e_0} - \beta + 1}}{1 - \left(\frac{q_t}{q_0}\right)^{\frac{1-e_0}{\eta_{11}e_0} - \beta}} q_0 \quad (4.57)$$

We are now ready to apply the theory and the models to some systems.

Problem 4.18
Show that if $\eta_{10} > 0$, then there is an infinite accumulation of carbon for all values of β.

4.10 Comparison with other approaches

There are a number of other approaches to describe carbon and nutrient dynamics in litters and soil organic matter (see review by Paustian et al. 1996). Most of the litter decomposition models describe the litter as a homogeneous substrate but with a changing specific decomposition rate. In none of these other models is there any explicit reference to properties of the decomposer community. Moreover, many of these models can be derived as special cases of (4.44) and (4.46) in the form of power models, $C(t) \sim t^{-k}$, where k defines the order of the assumed kinetics. Nor are these models extended to soil organic matter and most of them describe only mass loss. The soil organic matter models (e.g. CENTURY, Parton et al. 1987, The Rothamsted model, Jenkinson & Rayner 1977), on the other hand, are generally driven by a decomposer community and very much have the appearance of discrete approximations of the quality continuum. Most of these models can be derived as special cases of (4.7) (Bosatta & Ågren 1995a, Ågren et al. 1995) by appropriate choices of the decomposer functions.

NOTES

[1] This Chapter summarises much of the work in Bosatta & Ågren (1985, 1991ab, 1994) and Ågren & Bosatta (1987).

[2] e.g. Minderman 1968, Fogel & Cromack 1977, Jenkinson & Rayner 1977, van Veen & Paul 1981, Melillo et al. 1982, Berg & Ågren 1984, Berg et al. 1984, McClaugherty et al. 1985, McClaugherty & Berg 1987, Parton et al. 1987, Cheshire et al. 1988, White et al. 1988, Taylor et al. 1989, Ågren & Bosatta 1996. For single litters the interest has been focused around the concentration of sugars, polymer carbohydrates, lignin (including phenolic decomposition products) and nitrogen. These substances, in isolation or in different combinations, have been used as indexes of substrate quality. A general tendency towards the increasing importance of lignin with increasing degree of decomposition is found in these studies although the decomposition rate sometimes is better described by variables not including lignin.

[3] Work in this direction is also found in Aber & Melillo (1982) and Berg & Ågren (1984).

[4] Bosatta & Ågren (1995a) give further examples.

[5] Bosatta & Ågren (1985) derived and solved this equation under the assumption that $\eta_1(q)$ is constant.

[6] This pattern is often observed (e.g. Aber & Melillo 1980, Berg & McClaugherty 1989).

[7] In mathematical terms, (4.49) can be expressed as an incomplete gamma function (e.g. Abramowitz & Stegun 1972). We have, however, not found any use for its known mathematical properties.

[8] If $D(q',q)$ is given by a log-normal function, $D(q',q) = (2q'^2 s^2 \pi)^{-1/2} \exp\{-[\ln(q') - \ln(\varepsilon q)]^2/(2s^2)\}$, we get $\eta_1(q) = q(1 - \varepsilon \exp\{s^2/2\})$, i.e. a linear function in q.

5

Carbon and nitrogen - applications

We will now start comparing the predictions from the equations of Chapter 4 with experimental data on C and N dynamics for some simple conditions. Both models for decomposer functions will be used to show some of their specific features. The simplest systems to study are probably the incubation experiments of decomposition where a given, initial, amount of material is followed over time; these experiments are suited to being described by the equations corresponding to single litter cohorts. From such experiments we will obtain estimates of parameters defined in Chapter 4. Equations (4.47) and (4.48) describe the dynamics of C and N in the soil organic matter. These equations will be used to estimate the amount of the elements in the steady state of the soil. We compare these results with the experimental steady state amounts of C and N in different organic matter fractions of some particular ecosystems. We will also take a closer look at estimates of decomposer biomass.

5.1   Single litter cohorts

In Chapter 4 we proposed two different sets of models for decomposer properties. Before comparing predictions with empirical information we need to choose which models to use. Comparisons with the empirical information provide little help because all the models suggested are sufficiently flexible to give similar predictions if suitable parameters are included. This problem is made worse by the non-linear character of many functions and with strongly correlated parameters, which leads to an additional uncertainty in the choice of parameter values; there is an infinite combination of parameters that fit the predictions to data equally well. We will therefore use both sets of models with their associated parameters although, when recalling the arguments presented in section 4.8, our preferences are for Model II. This area will look for experiments that are specifically designed to discriminate between different decomposer models. In spite of this ambiguity, we want to emphasis that any one of the models and parameter sets can describe a wide range of conditions.

Leaf litter is an important category of litter entering the soil system. Decomposition of leaf litter has been extensively studied for many species and under many differing conditions. We will use one of those experiments with Scots pine needles (Berg & Ågren 1984) to calculate characteristic values for the parameters appearing in (4.44) and (4.46) (q_0 and the decomposer parameters η_{10}, η_{11}, f_C, e_0, e_1, β and u_0). A non-linear least-square regression method was used to fit equation (4.44) with q substituted with (4.46) to the data points for Model I. Table 5.1 gives the estimated parameters and Figure 5.1 shows the fit of the model to the data set. Since e_0 was set to 0 and the other parameters satisfy $\beta\eta_{10}e_1 < 1$, a steady state in the soil organic matter will be possible, cf. 4.7. For Model II the parameters η_{11}, e_0, and β were given the values estimated by Ågren & Bosatta (1996), see also Chapter 12, whereas q_0 and u_0 were adjusted to give the best fit. This parameter set, that we will use henceforth, is also given in Table 5.1, and Figure 5.1 shows the fit to data. Again, this parameter combination permits soil carbon and nitrogen to be in steady state. Both regressions explain about 95% of the variance and discrimination between the models on this ground is therefore not possible. The theory and its models can, at least, explain the mass losses of a single litter cohort. Throughout the remainder of this book these particular combinations of parameters will serve as a reference set for applications to other systems.

A more complicated test can be made by looking at the changes in carbon and nitrogen simultaneously. If $q \cong q_0$, we can use the Maclaurin expansion $\exp\{(q - q_0)/(e_1 q_0 \eta_{10})\} \cong [1 + (q - q_0)/q_0)]^{1/(e_1 \eta_{10})}$ to combine (4.44) and (4.45) into an expression relating the mass (carbon) loss to the nitrogen:carbon ratio. When we combine (4.50) and (4.51) we get the same result without any approximation

$$\ln g = \frac{1 - e(q_0)}{e(q_0)} \ln \frac{f_N/f_C - r}{f_N/f_C - r_0} \qquad (5.1)$$

where $e(q_0) = e_1 q_0$ for Model I and e_0 for Model II. McClaugherty et al. (1985) followed dry weight and N concentration for six litter types from different stands but incubated in the same stand for two years. The free

Table 5.1 *Parameters for decomposition of Scots pine needles. Frequently encountered combinations of parameters are also given. Note that with Model I we cannot get an absolute value for all parameters but sometimes only products or ratios.*

Parameter	Model I Value	Model II Value	Unit
f_C	0.5	0.50	—
q_0/η_{10}	1.12	—	—
q_0	—	1.10	—
η_{11}	—	0.36	—
β	3	7	—
e_0	0	0.25	—
$e_1 q_0$	0.19	0	—
$1/(\eta_{10} e_1)$	6	—	—
$(1 - e_0)/(e_0 \eta_{11})$	—	8.33	—
u_0	—	0.168	yr^{-1}
$(\beta - 1)\eta_{10} f_C u_0 q_0^{\beta-1}$	0.168	—	yr^{-1}
$\beta \eta_{11} f_C u_0 q_0^{\beta}$	—	0.418	yr^{-1}
$u_0 q_0^{\beta}$	0.188	0.332	yr^{-1}

parameters when fitting (5.1) to the data are $e(q_0)$ and f_N/f_C. It turns out that the goodness of fit does not change as long as $e(q_0)f_N/f_C$ is constant. With Model I we can therefore choose a fixed value of f_N/f_C ($= 0.1$) and estimate different values of $e(q_0)$. With Model II, where $e(q_0)$ is constant, we instead get different values for f_N/f_C. Figure 5.2 shows the data together with regression lines of the form (5.1). From the slope of the lines we calculate factors of $e_1 q_0$ from 0.19 (white pine needles) to 0.367 (aspen leaves) for Model I, and for Model II we get f_N/f_C between 0.146 (aspen leaves) and 0.070 (red maple wood chips). In all cases the coefficients of determination are larger than 0.95. These results for Model II suggest that f_N/f_C is not a unique decomposer property but depends on the substrate being decomposed. The values for f_N/f_C obtained when fitting with Model II correlate positively with r_0, i.e. when substrates of high initial nitrogen concentration are decomposed the decomposers also have high nitrogen concentration. If this is a result of decomposers attacking different materials, or if it is a result changing nitrogen concentration with substrate is an open question. We have, however, to be aware that this might mean that f_N/f_C cannot be set constant over arbitrarily long ranges of decomposition.

We can also use (5.1) in another way with Model I. A basic requirement in the development of the theory has been that we can identify parameters uniquely either with substrate (litter) properties or with decomposer properties. This requirement is fundamental because without it the possibilities of interpreting the components of the theory decrease drastically. Now, e_1 is a decomposer and q_0 a substrate property. From (5.1) we can calculate e_1q_0 as done above. Suppose now that we have different decomposers i and j and different substrates x and y. Then the following relation obtains between estimates of e_1q_0.

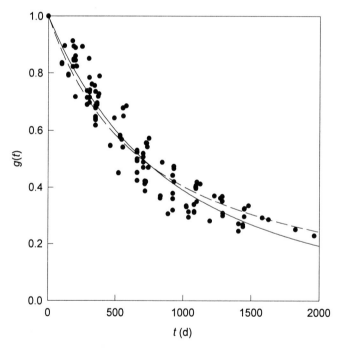

Figure 5.1 Mass of decomposing Scots pine needles relative to initial mass versus time. Solid line: Model I. Broken line: Model II (modified from Ågren & Bosatta 1987).

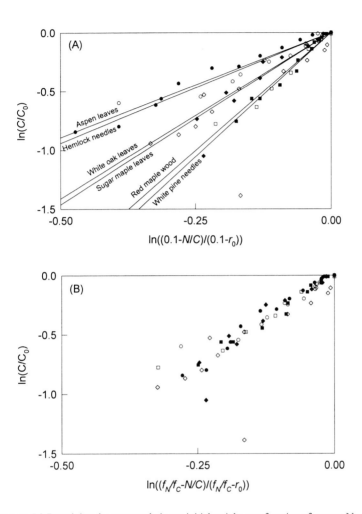

Figure 5.2 Remaining dry matter relative to initial weight as a function of current N concentration (transformed) for six different litter types (McClaugherty et al. 1985) and calculated regression lines (some obvious outliers have not been included in the regressions): (O) = hemlock needles. (●) = aspen leaves. (◇) = sugar maple leaves. (◆) = white oak leaves. (□) = red maple wood chips. (■) = white pine needles. (A) Model I (from Bosatta & Ågren 1985). (B) Model II.

Table 5.2 *Estimated decomposer efficiencies,* e_1q_0, *for six combinations of fungi and substrates (from Lekkerkerk et al. 1990).*

	P. oleracea	P. chrysosporium
Birch	0.257	0.147
Aspen	0.178	0.116
Spruce	0.145	0.099

$$e_{1i}q_{0x}e_{1j}q_{0y} = e_{1i}q_{0y}e_{1j}q_{0x} \tag{5.2}$$

Equation (5.2) was tested by Lekkerkerk et al. (1990) using two different fungi (*Phanerochaete chrysosporium* Burdsall and *Poria oleracea* Davidson & Lombard) and wood from three different tree species (birch, *Betula pubescens* Ehrh., aspen, *Populus tremula* L., and Norway spruce, *Picea abies* Karst.). Table 5.2 shows the estimates of the initial decomposer efficiency, e_1q_0. As is clear from the table for all wood species, *P. oleracea* is more efficient than *P. chrysosporium* and for both fungi the initial qualities rank as birch>aspen>spruce. A more firm statistical analysis shows that it is not possible to reject the hypothesis stated in (5.2). This experiment, therefore, supports the parameterisation of the efficiency function, $e(q)$. There is, however, an uncertainty in the estimates of f_N/f_C such that, in consistency with Model II, it would be possible to explain the results in terms of a variability in f_N/f_C instead.

Among other things, mass losses and changes in nitrogen were studied for two years by Aber et al. (1984) using twelve litter types in five different forests. They observed different patterns of N accumulation and release among litter types but virtually all litters accumulated N during some part of their decomposition. We want to contrast their measured values of the "nitrogen factor" with the theoretical values calculated with the use of (4.50) and (4.51).

The expression for the maximum amount of nitrogen in the litter, h_c, can be derived for Model I with the same approximation as used when deriving (5.1) and by assuming that $r_c \cong e_1q_0f_N/f_C$

$$h_c = \frac{e_1q_0f_N}{f_C}\left[\frac{1-e_1q_0}{1-r_0/(f_N/f_C)}\right]^{(1-e_1q_0)/e_1q_0} \tag{5.3}$$

which is almost identical to (3.19) for the homogeneous litter; this is not surprising because it has been assumed that only small changes in litter quality
have occurred up to q_c ($q_c \cong q_0$) so the decomposer efficiency remains at its
initial value.

For 58 data sets and with the values of $e_1 q_0$ obtained according to (5.1) we
have calculated the nitrogen factor, h_c, and compared it with the experimentally observed ratios between the maximum N content in the substrate and
initial N amount. Figure 5.3 shows these results. The two ways of obtaining
this ratio agree very well with each other, except for three points having the
highest ratios. However, these three points are for the data sets of white pine
wood which, after 2 years of decomposition have still not reached their

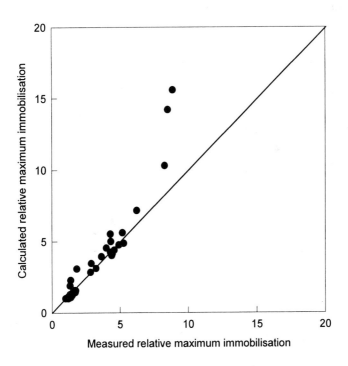

Figure 5.3 Theoretically calculated ratio between nitrogen amount in the substrate at the critical
point and the initial amount versus the ratio between the observed maximum nitrogen amount
and initial nitrogen amount (from Bosatta & Ågren 1985).

critical N:C ratio. Hence, these substrates will continue to accumulate N and the corresponding points in Figure 5.3 will move to the right, decreasing the discrepancy between theoretical and experimental values.

Problem 5.1

Derive (5.1) for model I.

Problem 5.2

Derive (5.3) for Model I and II.

Problem 5.3

Using (4.33), calculate the general equation that defines the critical n:C ratio with Model I. Compare it with the approximation $r_{nc} \cong f_n e_1 q_0 / f_C$ used above.

Problem 5.4

Derive equation (5.1) and compare with the results of Problem 4.13.

5.2   Steady state of soil organic matter

The existence of a steady state depends critically upon the set of parameters. For example, a steady state with finite amounts does not exist with the conditions of Figure 4.4 (A) since each litter cohort leaves behind a finite remainder (the integrand is always positive and diverges, so does the integral). On the other hand, under the conditions of Figure 4.4 (B) a steady state with finite amounts of C and N can be possible; this obtains if $\beta \eta_{10} e_1 < 1$ (case (ii) and (iii) in the figure). With $e_0 = 0$, we have the term $q^{1/\eta_{10} e_1 - \beta}$ in the integrand (4.49), which is convergent for $q = 0$ only if the previous condition is satisfied. This condition also means that, as discussed in section 4.6, the degradation in quantity goes faster than the degradation in quality, thereby enabling the existence of the steady state. For Model II, the requirements for steady state are ($\eta_{10} = e_1 = 0$) $\beta \eta_{11} e_0 < 1 - e_0$. The steady state soil stores of carbon and nitrogen for constant rates of litter input are easily derived from (4.53) and (4.54) and are shown in Figure 5.4.

We can extend the use of the parameters for Scots pine needles to any litter fractions by giving the initial litter quality relative to the initial quality of pine needles. For example, the factor $u_0 q_0^\beta$ is substituted with $u_0 q_{0,pn}^\beta (q_0/q_{0,pn})^\beta$ where $q_{0,pn}$ is the initial quality of pine needles. The sensitivity of some variables to properties of the initial litter (r_0 and q_0) are shown in Figure 5.4. The initial specific decomposition rate varies rather evenly, although rapidly, over an expected range of initial litter qualities. The steady state carbon and nitrogen stores in the soil also change smoothly but in the opposite direction.

Initial nitrogen concentrations seem to be of minor importance unless they are associated with changes in f_N/f_C, which might give rise to much larger effects. In particular, if low qualities also are correlated with low nitrogen concentrations, as is likely to be the case, we should see nitrogen stores that are insensitive to litter quality.

Problem 5.5

Using the decomposer functions $u(q) = u_0$ and $e(q) = e_0$ where u_0 and e_0 are two positive constants, calculate (4.47) and (4.48).

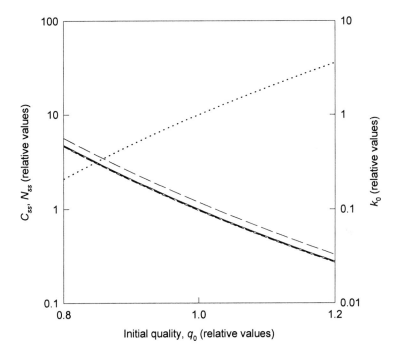

Figure 5.4 The steady state soil stores of carbon, C_{ss} from equation (4.53), and nitrogen, N_{ss}(broken lines) from equation (4.54) (solid line), and initial specific decomposition rate for the litter, k_0, (dotted line) as a function of initial litter quality and initial nitrogen concentration ($r_0 = 0.05$, 0.005, 0.0025; for $r_0 = 0.010$ the line coincides with the carbon line). Both stores and litter qualities are given relative to Scots pine needle litter ($r_0 = 0.010$). $f_N/f_C = 0.10$ (modified from Ågren & Bosatta 1987).

5.3   Correlation between carbon and nitrogen turnover

Combining (4.53) and (4.54) under steady state conditions, we get

$$\frac{N^{ss}}{I_0 r_0} = \left[\frac{f_N / f_C}{r_0} + (1 - \frac{f_N / f_C}{r_0}) \frac{1 - e_0 - \beta \eta_{11} e_0}{1 - \beta \eta_{11} e_1} \right] \frac{C^{ss}}{I_0} \tag{5.4}$$

as the relation between the turnover times of carbon and nitrogen.

Equation (5.4) suggests that a linear relationship should exist between the turnover of the two elements. Furthermore, using the previous parameter values, the proportionality factor can be calculated to lie between 1 and 10 for the different litter fractions, implying that the turnover times of nitrogen are clearly longer than those of carbon, cf. Table 2.2. This reflects the fact that nitrogen, in contrast to carbon, is recycled in the soil.

We have used data collected on stores and inputs of carbon and nitrogen from a number of different terrestrial ecosystems by Rodin & Bazilevich (1967) and Cole & Rapp (1981) to calculate the turnover times of carbon and nitrogen in a number of ecosystems. Figure 5.5[1] shows these data together with the regression line which is

$$\ln \frac{N^{ss}}{I_0 r_0} = 0.31 + 1.00 \ln \frac{C^{ss}}{I_0}, \quad r^2 = 0.95, \quad n = 42 \tag{5.5}$$

The intercept of the regression line is in the range predicted from (5.4); $e^{0.31} = 1.4$. To establish a strict correspondence between (5.4) and (5.5) we must assume that (5.4), which applies to each litter fraction separately, can be made to apply to the whole mixture of litter fractions, viz., that q_0 and r_0 in (5.4) are representative of some kind of mean litter fraction.

5.4   Steady state of a Scots pine forest

Equations (4.49) and the corresponding equation for nitrogen were used by Ågren & Bosatta (1987) in order to calculate the amounts of carbon and nitrogen in the soil of an old Scots pine forest.[2] Four litter fractions were considered: canopy, tree roots, stems and ground vegetation. All litter formed from the tree canopy was lumped and treated as needles; the contribution from stem-wood was estimated as being equal to the current rate of stem-wood production and all ground vegetation was treated as one component.

Fluxes and stores of carbon and nitrogen were taken from Andersson et al. (1980).

The decomposer parameters were assumed to be common to all litter fractions. On the other hand, the initial qualities of the four litter fractions were adjusted relative to the q_0-value of needles (reference quality) in such a way that the initial decomposition rates are correct according to (4.10) - it should be pointed out that the value of the parameter β is critical in this context. The initial decomposition rates have been set to (roots and ground vegetation): $k_0 = 0.16$ yr^{-1} (Berg 1984) and (stems) $k_0 = 0.028$ yr^{-1} (Barber & Van Lear 1984; Sollins et al. 1980). The decomposition rate of ground vegetation was assumed equal to that of Scots pine roots. Finally, f_N/f_C was set to 0.1.[3] This gives soil carbon and nitrogen contents as shown in Table 5.2.

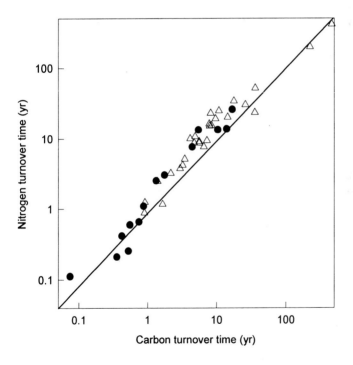

Figure 5.5 Turnover time of soil nitrogen versus turnover time for soil carbon for different ecosystems. Data from Rodin & Bazilevich (1967) = ●; Cole & Rapp (1981) = △ (from Ågren & Bosatta 1987).

Table 5.3 *Estimates of soil carbon and nitrogen amounts in an old Scots pine forest.* $q_{0,c}$ *is the q_0-value for the canopy fraction.*

Litter fraction	I_{0L} kg ha^{-1}yr^{-1}	k_{0L} yr^{-1}	$q_{0L}/q_{0,c}$ —	r_{0L} kg(kg)$^{-1}$	C_L^{ss} kg ha^{-1}	N_L^{ss} kg ha^{-1}
Canopy	710	0.23	1	0.010	4490	142
Tree roots	950	0.16	0.817	0.015	9068	285
Stems	750	0.028	0.325	0.0026	40296	433
Ground vegetation	975	0.16	0.817	0.0076	9306	240
Total	3385				63160	1100
Observed					41000	1035

The contribution to the soil carbon and nitrogen from stem-wood litter is significant, 64 and 39%, respectively. Yet, the carbon and nitrogen input from this source is much lower, 22 and 6% of the total litter production, respectively. This means that sources of low initial quality give a high contribution to the soil organic matter. The accuracy in the estimates of soil carbon and nitrogen content are therefore critically dependent upon correct estimates of the decomposition rate of these recalcitrant fractions or, equivalently, on correct estimates of its initial quality.

Small numerical changes in other parameter values (especially in β) could improve the agreement between calculated and observed soil carbon and nitrogen contents in Table 5.2. However, in view of the precision of the other parameter estimates, as well as the unknown importance of factors like forest fires, it does not seem meaningful to try to further close the discrepancies between observed and calculated values. A better description of the variation in inputs of litter types with stand age is also desirable. The point is, however, that we can show that the predictions of our theory give the correct orders of magnitude.

5.5 An agricultural application

Long-term changes in soil carbon and nitrogen resulting from different regimes of fertilisation and organic matter addition were studied in the field in the "Ultuna continuous organic matter experiment", established in 1956 in central Sweden.[4] Hyvönen et al. (1996) used Model II to analyse what happened to plots where six types of organic matter were added

Table 5.4 Parameters and results for the analysis of the Ultuna continuous organic matter experiment. All units for amounts g m^{-2}. $u_0 = 3.1\ 10^{-4}\ d^{-1}$ in all treatments.

Treatment	q_0	r_0	f_N/f_C	Added 1956-1991		Measured remaining 1991		Simulated remaining 1991	
				C	N	C	N	C	N
Green manure	1.15	0.056	0.18	6710	370	920	90	920	90
Yellow straw	1.09	0.014	0.18	6700	100	1220	80	1210	90
Sawdust	0.99	0.0016	0.07	6760	10	1990	40	1960	40
Farmyard manure	0.98	0.047	0.18	6610	310	1960	160	1910	160
Sewage sludge	0.83	0.11	0.07	6480	710	3590	340	3550	370
Peat	0.75	0.01	0.07	6790	110	4680	90	4670	100

biannually: sewage sludge, green manure, farmyard manure, yellow straw, peat and sawdust. The plots were cropped sequentially with barley (*Hordeum vulgare* L.), oats (*Avena sativa* L.), beets (*Brassica* spp.) and occasionally rape (*Brassica napus* L., var. *oleifera*). All harvests, including straw, were exported from the field, leaving only stubble (< 2 cm), spillage and roots.

The parameter estimations along with the results are presented in Table 5.3 and Figure 5.6. It is clear that when parameters are used that are consistent with those used in the other applications, the observations will be reproduced well in spite of the very different characteristics of the substrates.

5.6 Decomposer biomass

It has been suggested to use decomposer biomass as an indicator of state and change of soil organic matter because the decomposers respond much more rapidly than the soil organic matter as a whole to changes in the management, climate etc.[5] Let us now examine how this can be done. From equation (4.4) we can calculate the carbon content in the decomposer biomass. Assuming a time and quality-independent mortality rate, $\mu_b(q,t) = \mu$, we get the total microbial carbon, $C_b(t)$, by integrating $\rho_b(q,t)$ over q

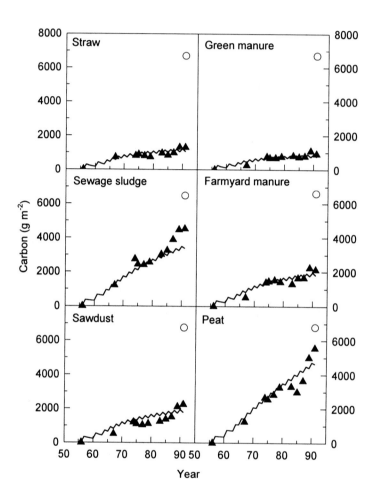

Figure 5.6 Development of carbon and nitrogen from additions of different types of organic substrates. Predicted values = lines; measured values (▲); total additions (○) (modified from Hyvönen et al. 1996).

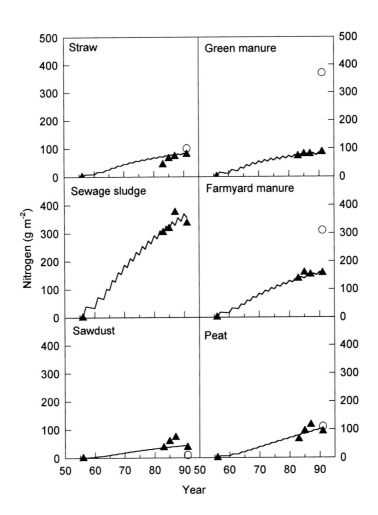

$$C_b(t) = \int_0^\infty \frac{1}{\mu_b(q,t)} \left[\int_0^\infty f_C D(q,q')u(q')\rho(q',t)dq' \right] dq = \frac{f_C}{\mu} \overline{u(q_t)}C(t)$$

$$(5.6)$$

or with the usual (4.37) as model for decomposer growth rate

$$\frac{C_b(t)}{C(t)} = \frac{f_C}{\mu} u_0 \overline{q_t^\beta} \tag{5.7}$$

Note that (5.7) requires the mean of q^β, not the mean of q raised to the power of β. However, for a single litter cohort the two are exchangeable. With Model II, we can calculate the mean in (5.7) for a system with a constant litter input rate, to obtain

$$\frac{C_b(t)}{C(t)} = \frac{f_C u_0 q_0^\beta}{\mu} \frac{1-e_0-\beta e_0 \eta_{11}}{1-e_0} \frac{1-(q_t/q_0)^{(1-e_0)/(e_0 \eta_{11})}}{1-(q_t/q_0)^{(1-e_0)/(e_0 \eta_{11})-\beta}} \tag{5.8}$$

Equation (5.7) is exact, but the $g(q)$ from (4.50) used to calculate (5.8) is based on the first-order moment expansion. With the parameters in Table 5.1, this means that at steady state the decomposer biomass relative to the soil carbon is a quarter of that found on fresh litter. Bosatta & Ågren (1994) studied some of the implications of this equation for different soils at different ages after perturbations.

Anderson & Domsch (1990) estimated μ to be between $1 \cdot 10^{-4}$ and $3 \cdot 10^{-4}$ h^{-1} at 22 °C under different agricultural practices. With the higher mortality rate, which we will use below, and parameters from Table 5.1, but u_0 increased with a factor 4 to account for the higher temperature, we get $C_b/C = 0.03$ which agrees with the ordinarily assumed values of 0.01-0.05.

Problem 5.6
Derive (5.8).

If we want to follow how the decomposer biomass changes as a litter ages we can eliminate t (or rather q_t) between (5.7) and (4.50) to get an expression for C_b/C as a function of remaining substrate, g:

$$\frac{C_b}{C} = \frac{f_C u_0 q_0^\beta}{\mu} \left(\frac{C}{C_0} \right)^{e_0 \eta_{11} \beta/(1-e_0)} \tag{5.9}$$

Figure 5.7 provides a comparison of (5.9) with some empirical studies on wheat straw (Bottner et al. 1984), ryegrass (Jenkinson & Rayner 1977, Cerri & Jenkinson 1981), and *Medicago litoralis* (Ladd et al. 1981). We have used the q_0-value given by Ågren & Bosatta (1995) for wheat straw and a somewhat higher value to show what happens with higher qualities. All other parameters are taken from Table 5.1. We fail to predict the first part of the decomposer biomass curve for ryegrass. Since (5.9) is based on the adiabatic assumption that the decomposer biomass is in equilibrium with its substrate, we expect to underestimate the decomposer biomass in an initial transient phase, as is seen in the figure.

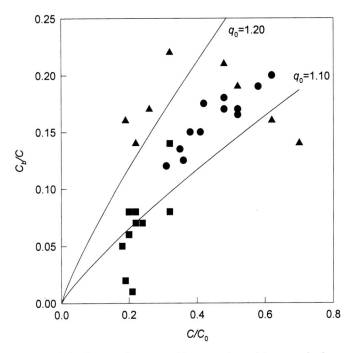

Figure 5.7 Relation between decomposer biomass and remaining mass in decomposing litter cohorts. Solid lines are theoretical predictions for three substrates of different initial qualities: (●) wheat straw, (▲) ryegrass, and (■) *Medicago litoralis*. The decomposer mortality has been been set to $0.75 \cdot 10^{-4}$ h^{-1} to correct for a lower temperature in the field than in the laboratory (from Bosatta & Ågren 1994).

Problem 5.7
Derive (5.9).

NOTES

[1] This figure is similar to Figure 1.18 in Swift et al. (1979).

[2] See Axelsson & Bråkenhielm (1980) for detailed site description.

[3] This value is adopted from Bosatta & Staaf (1982).

[4] Location 59°48'N, 17°38'E, see Persson (1980) and Kirchmann et al. (1994) for details.

[5] Powlson et al. (1987), Insam & Domsch (1988), Anderson & Domsch (1989), Insam et al. (1989), Wolters & Joergensen (1991, Sparling (1992).

6

Carbon, nitrogen, phosphorus and sulphur - applications

In this chapter the interactions between the cycles of C, N, P, and S in decomposing litters and soils are analysed. The theoretical predictions are compared against empirical data on C, N, P and S behaviour in a number of different conditions covering both steady states of soils and dynamics in incubation experiments in laboratories. The agreement between theoretical and empirical results indicates that the cycles of N, P, and S basically follow similar rules. We will use Model II (see section 4.9) for decomposer properties in this chapter.

6.1   C-N-P-S interactions

The importance of interactions between nutrient cycles in soils has long been recognised. Significant similarities between nutrient cycles have been identified. Carbon, N, P, and S are apparently stabilised together in soil and an increase or decrease in one of the elements is accompanied by a parallel increase or decrease in the others.[1] Such a relationship suggests that the measure of biological stability exhibited by soil organic matter must be achieved through reactions of large organic aggregates or molecules containing C, N, P and S in approximately the same proportions as these elements are found in microbial tissues.[2] In spite of the central role of S and P, these nutrients have received only limited attention from a theoretical or modelling point of view.[3]

Many empirical data indicate that the stabilisation and dynamics of the four elements cannot be explained only in terms of decomposer properties or in terms of the nutrient:C ratio in the substrate. We want to show how long-term behaviour of P and S in litters and soils can be conveniently explained in terms similar to those used for N if quality is taken into consideration (see Chapter 5). Our analysis is based entirely on the properties of the soil organic matter. Soil properties like texture and mineral P availability, which are important for stabilisation of soil C (Stewart, 1984), are not explicitly represented. Certain aspects can easily be accounted for through the parameters of the theory. For example, increasing the parameter β in (4.37) will slow down the decomposition of the most recalcitrant material and could be used to simulate the effects of a physical protection by an increasing clay content. When the decomposition products are reacting chemically with soil P, it might be better to consider the consequences as resulting from changes in the dispersion function (the parameter η_{10} or the function $D(q,q')$ in (4.7)). In a situation where soil properties are promoting chemical reactions between decomposition products it might be necessary to add additional, non-linear, terms to (4.7).

6.2   C, P and N dynamics in single litter cohorts

Berg & McClaugherty (1989) followed the dynamics of mass, N and P in leaf litters from 11 different species incubated in 5 different forest soils and in the laboratory. Figure 6.1 shows the results from one of these incubations together with the corresponding theoretical relation from (4.50) and (4.51). The parameters are from Table 5.1 for Model II and $f_N/f_C = 0.033$. Swift et al. (1979) estimated a range of 0.06 to 0.12 for the P:N ratio of fungi and bacteria growing in different natural substrates; we used a value of $f_P/f_C = 0.002$ ($f_P/f_N = 0.061$). The initial N:C and P:C ratio of pine needles, r_{0N} and r_{0P}, were set to the measured values.

Nitrogen and P develop similarly, passing over maximum amounts. With Model II we get from (5.3) the following equation (exactly when e_1q_0 is replaced by e_0) for the ratio between the maximum amounts of P and N in a single litter cohort (see also Problem 5.2)

$$\frac{h_{cP}}{h_{cN}} = \frac{f_P}{f_N} \left[\frac{1 - f_C r_{0N} / f_N}{1 - f_C r_{0P} / f_P} \right]^{\frac{1-e_0}{e_0}} \tag{6.1}$$

The ratio depends only on initial nutrient:C ratios of the litter cohort relative to the decomposer nutrient:C ratios. Different litter materials show a great variability in the values of r_{0N} and r_{0P} and considerable variations can be found even within materials of the same quality. As regards all the species and growth conditions in the study of Berg & McClaugherty (1989), Figure 6.2 shows that the variability in r_{0N} and r_{0P} can reasonably well explain the variability in the ratio defined in (6.1).

Problem 6.1

Derive equation (6.1)

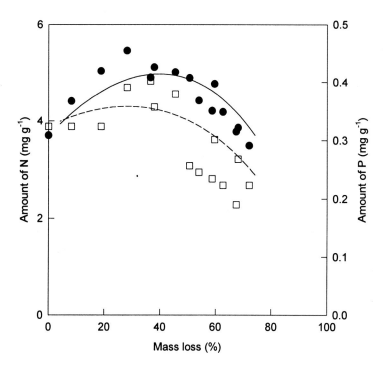

Figure 6.1 Amounts of N (● and solid line) and P (□ and broken line) versus mass loss in Scots pine needles incubated in a forest soil (Berg & McClaugherty 1989). The theoretical curves are calculated with (4.50) and (4.51) (redrawn from Bosatta & Ågren 1991b).

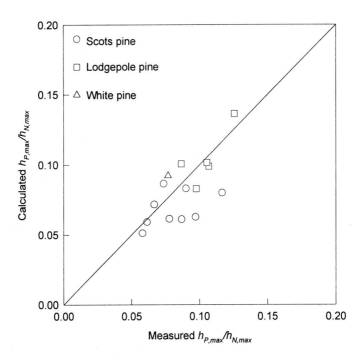

Figure 6.2 Theoretically calculated ratio between maximum amount of N and maximum amount of P, equation (6.1), versus the ratio between the observed maximum amount of P and the observed maximum amount of N in needles of three pine species incubated in the field and the laboratory. Data from Berg & McClaugherty (1989). (redrawn from Bosatta & Ågren 1991b).

6.3   Stabilisation of C, N, P and S in the soil

Let $C_s = \Sigma_L C_L^{ss}$ and $n_s = \Sigma_L n_L^{ss}$ denote the total amounts of C and nutrient n at steady state in soil that is made up of several different litter types L. Using (4.53) and (4.54), several relations between C_s, N_s, P_s and S_s can be derived; the following relates the amounts of C, N and S in a soil at steady state

$$N_s = \frac{f_N}{f_S} S_s + C_s \sum_L \left(r_{0NL} - \frac{f_N}{f_S} r_{0SL} \right) \left(1 - \frac{e_0}{1 - \beta \eta_{11} e_0} \right) \frac{C_L^{ss}}{C_s} \qquad (6.2)$$

Substituting S_s with P_s, f_S with f_P and r_{0SL} with r_{0PL}, (6.2) becomes an equation relating N to P and C in steady state. Equation (6.2) suggests that if N_s is plotted against S_s (or P_s) the slope is going to be proportional to f_N/f_S (or to

f_N/f_P). Furthermore, the absolute value of the intercept must be proportional to C_s but the signs of $r_{0NL} - (f_N/f_S)r_{0SL}$ and $r_{0NL} - (f_N/f_P)r_{0PL}$ can be either positive or negative, depending on the relationship between the P, S and N concentrations in litters and decomposers.

Figure 6.3 shows the regression of N_s on S_s for a large number of soil types from New Zealand and Sweden. The concentrations of C in the Swedish soils are, with a few exceptions, one order of magnitude (~ 30 g/kg soil against $\sim$ 3 g/kg soil) larger than in the New Zealand soils (the four Swedish samples lying closest to the regression line from the New Zealand soils have markedly lower C concentrations). Values of f_N/f_S have been estimated to be in the range 10 to 20 (Saggar et al. 1981), which agrees with the slopes in Figure 6.3. Equation (6.2) also suggests that if the regression of S_s /C_s on N_s/ C_s is analysed instead, all points of Figure 6.3 should fall along a line with a slope of f_S/f_N (Figure 6.4).

In another study by Thompson et al. (1954), the relation between N and P in a large population of virgin and cultivated soils spanned a wide variety of soil types. Virgin and cultivated soils had the same slopes but the intercept was approximately twice as large in the virgin soils when N concentration was plotted against P concentration (Figure 2 in the cited reference). Consistent with (6.2), there was twice as much C in the virgin soils than in the cultivated soils.

Figure 6.5 shows a similar regression of P_s/C_s on N_s/C_s using the data from Thompson et al. (1954) (soils with pH < 7.0) and Pearson & Simonson (1939) on virgin and cultivated soils in Iowa; the slope falls within the estimated range of values of f_P/f_N (see above). Data from Texas and Colorado soils (Thompson et al. 1954) do not fit well to the same regression line. They fall consistently well below it and, as in Figure 6.3, are associated with lower C concentrations.

Problem 6.2
Derive equation (6.2).

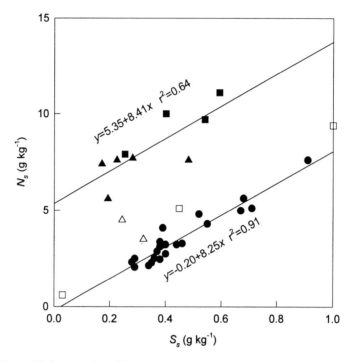

Figure 6.3 Concentration of N versus concentration of S in a number of different soils. Data from Walker & Adams (1958) on virgin New Zealand soils (●) and Nömmik et al. (1984) on Swedish forest soils (other symbols). The regression lines are calculated for each soil separately. Points with open symbols are not included in the calculation of the regression for the Swedish soils (redrawn from Bosatta & Ågren 1991b).

It may be argued that the dispersion in the data of Figure 6.5 could be decreased by considering some details of the chemistry of soil P. For example, soils having pH > 7 exhibit a higher N_s/P_s ratio than soils having pH < 7 (Thompson et al. 1954) which may require extra terms in Eq. (6.2) for its explanation. There is, however, an alternative way of explaining the dispersion of Figure 6.5. Since the availability of phosphorus is sensitive to pH, with decreasing availability as pH increases above 6.5, we should expect a variability in r_{0P} relative to r_{0N}, which has to be accounted for before we consider soil effects on P mineralisation and immobilisation.

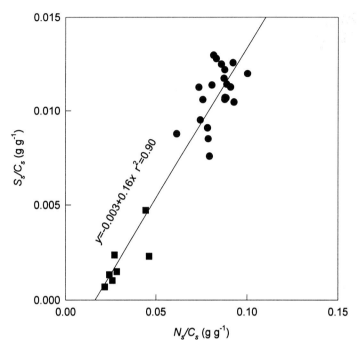

Figure 6.4 S:C ratio against N:C ratio. Data sources as in Figure 6.3 (redrawn from Bosatta & Ågren 1991b).

6.4   C, N, P and S mineralisation

Suppose that, at some time t_p, a sample of organic matter is taken. To calculate the evolution of the organic nutrients in the sample we must modify (4.48). The first effect of the perturbation is to interrupt the input of fresh material, thus the integrals extend only up to the oldest cohort t_p. The second effect is to change decomposer properties. We assume that the perturbation affects decomposer growth rate (under laboratory conditions it will normally be higher than under soil conditions) but that it does not affect decomposer efficiency or quality dispersion. To modify $u(q)$ means that $q(t)$ in the sample (and with it $g(q)$ and $h_n(q)$) will differ from $q(t)$ in the soil. More precisely, we assume that the perturbation affects only the level of the growth function, u_0.

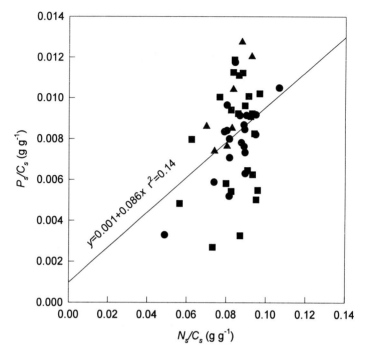

Figure 6.5 P:C ratio versus N:C ratio in a number of different soils. Data from Thompson et al. (1954) on virgin (●) and cultivated (■) soils and from Pearson & Simonson (1939) on upper layers of different types of Iowa soils (▲) (redrawn from Bosatta & Ågren 1991b).

The quality of a litter cohort that was of age a when the sample was taken will, at time $t > 0$ after the sample was taken, have quality

$$q(a,t) = q_0\left(1 + f_C\beta\eta_{11}q_0^{\beta}(u_0a + u_pt)\right)^{-1/\beta} \tag{6.3}$$

where u_p is the base decomposer growth rate after sampling (in the laboratory). Let us assume that the organic matter is at steady state when the sample is taken (i.e. that t_p is very large, such that $q(t_p,t)$ can be set to 0). We can then evaluate (4.48) as follows

$$n(t) = I_0 \int_0^{t_p} h_n(q_{a,t}) \, da =$$

$$= I_0 \int_{q_{0,t}}^{q_{t_p,t}} h_n(q) \frac{da}{dq} \, dq = \frac{I_0}{f_C u_0 \eta_{11}} \int_{q_{t_p,t}}^{q_{0,t}} h_n(q) q^{-\beta-1} \, dq =$$

$$= \frac{I_0 e_0}{f_C u_0 q_0^\beta} \left[\frac{f_n}{f_C} \frac{1}{1-e_0-\beta e_0 \eta_{11}} \left(\frac{q(0,t)}{q_0} \right)^{\frac{1-e_0}{e_0 \eta_{11}}-\beta} \right.$$

$$\left. - \left(\frac{f_n}{f_C} - r_{0n} \right) \frac{1}{1-\beta e_0 \eta_{11}} \left(\frac{q(0,t)}{q_0} \right)^{\frac{1}{e_0 \eta_{11}}-\beta} \right]$$

(6.4)

The fraction $\Phi(t)$ of original organic nutrient that has accumulated in inorganic form at time t after the perturbation is then

$$\Phi(t) = 1 - \frac{\dfrac{f_n}{f_C} \dfrac{1}{1-e_0-\beta e_0 \eta_{11}} \left(\dfrac{q(0,t)}{q_0} \right)^{\frac{1-e_0}{e_0 \eta_{11}}-\beta} - \left(\dfrac{f_n}{f_C} - r_{0n} \right) \dfrac{1}{1-\beta e_0 \eta_{11}} \left(\dfrac{q(0,t)}{q_0} \right)^{\frac{1}{e_0 \eta_{11}}-\beta}}{\dfrac{f_n}{f_C} \dfrac{1}{1-e_0-\beta e_0 \eta_{11}} - \left(\dfrac{f_n}{f_C} - r_{0n} \right) \dfrac{1}{1-\beta e_0 \eta_{11}}}$$

(6.5)

We have estimates of all parameters entering (6.5) except for u_p. Assuming a difference of 25 °C between laboratory and soil conditions together with a $Q_{10} = 2$, gives an approximately six-fold increase in decomposer activity. Thus, u_p is estimated to be $\cong 1$ yr^{-1} which is six times higher than the value 0.168 yr^{-1} estimated in Chapter 5.

Studying the N mineralisation of a large number of different soil types incubated in the laboratory at 35 °C, Stanford & Smith (1972) estimated that the fraction of N mineralised was proportional to $t^{1/2}$. Figure 6.6 is a plot of $\Phi(t)$ against t and $t^{1/2}$ showing that such a relationship is not unreasonable. The initial qualities in Figure 6.6, 0.9 to 1.1 (and Model II), cover a range of litter extending, in the case of forests, from canopy to woody litter (Table 5.1).

The asymptotic value of $\Phi(t)$, $\Phi(t = \infty)$, is 1 if $\beta \eta_{11} e_0 < 1$, (6.5); if instead the inequality sign is in the other direction, the asymptotic value of Φ is always less than 1 and a finite amount of organic nutrient remains forever in the substrate without mineralising.

Problem 6.3

Verify the conditions leading to complete or incomplete mineralisation defined in the previous paragraph.

Consider next a sample of soil organic matter made up of several litter types. The cumulative amount of nutrient mineralised at time t, $n_i(t)$, is formed by the contribution from all litter fractions L, and by definition of $\Phi_L(t)$

$$n_i(t) = \sum_L n_L^{ss} \Phi_L(t) \tag{6.6}$$

Equation (6.6) shows that, in general, $n_i(t)$ will not be proportional to n_s ($= \Sigma_L n_L^{ss}$), i.e. to the original amount of the nutrient in the sample. Contradictions in the literature on this point could be explained by (6.6). In soils with an input of homogeneous materials, $n_i(t)$ should be closely proportional to n_s. The contradictions can also depend on the time of incubation. As Figure 6.6 shows, the larger the incubation time, the larger are the differences between the cumulated mineralisation of different litter fractions.

Equation (6.6) suggests that if the organic matter of the soil is not too heterogeneous and the incubation times are not too large, we can expect some proportionality between $n_i(t)$ and n_s. Consequently, the amounts mineralised by two different nutrients are going to be approximately in the same relationship as their steady state values. This implies that if $S_i(t)$ or $P_i(t)$ are plotted against $N_i(t)$, the slopes should be close to f_S/f_N and f_P/f_N, respectively, even if the conditions of incubation are different. Thompson et al. (1954) measured P mineralised during 25 days in incubations at 40 °C from virgin and cultivated soils; Tabatabai & Al-Khafaji (1980) measured S mineralised from 12 soils at 35 °C during 26 weeks of incubation. Figure 6.7 shows the regression of P_i and S_i on N_i using these data. P_i and S_i apparently follow the same regression on N_i as expected when f_P/f_N and f_S/f_N are similar (both between 0.1 and 0.2).

Problem 6.4

Derive (6.4) and (6.5).

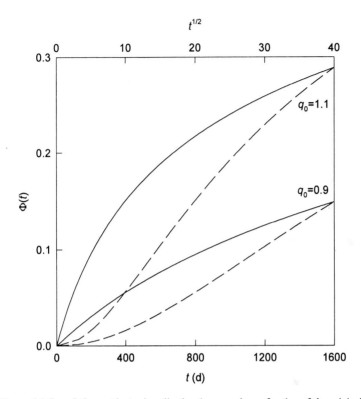

Figure 6.6 Cumulative nutrient mineralisation (expressed as a fraction of the original organic nutrient) as a function of time (solid lines) and the square root of time (broken lines) for two different initial qualities of the litter fractions (redrawn from Bosatta & Ågren 1991b).

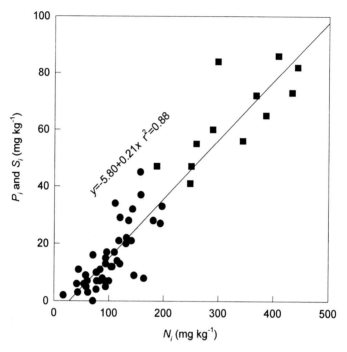

Figure 6.7 Phosphorus (●) and S (■) mineralised versus N mineralised for different soils. Data from Thompson et al. (1954) and Tabatabai & Al-Khafaji (1980) (redrawn from Bosatta & Ågren 1991b).

NOTES

[1] Thompson et al. (1948), Thompson & Black (1950), Thompson et al. (1954), Kaila (1949), Pearson & Simonson (1939), Schollenberger (1920), Walker & Adams (1958).
[2] Bartholomew (1965).
[3] The few existing models describe the S or P turnover using discrete organic matter fractions (box models) (Cole et al. 1977, Hunt et al. 1983, Parton et al. 1988) or are purely descriptive (Ellert & Bettany 1988).

7

Interactions with abiotic factors

In this Chapter we first analyse the dynamics of mineralisation-immobilisation in single litter cohorts when the availability of inorganic nitrogen is variable. This allows us to identify terms corresponding to net and gross mineralisation. These analyses are then extended to soils and the characteristics determining the immobilisation potential of inorganic nitrogen are calculated. Perturbations of the decomposer growth rate, exemplified by liming are discussed next. The Chapter is concluded with a generalisation of the perturbations discussed previously to other types of time-dependencies in the decomposer growth rate. Some technical aspects of the handling of the theory are also of importance.

7.1   The problems

In his pioneer research using ^{15}N techniques, Jansson (1958) established the importance of the mutually competing processes of mineralisation and immobilisation upon the fertility of soils. During the first half of the century, Doryland,[1] Waksman (1924) and others put forward a conceptual scheme to explain the behaviour of mineralisation-immobilisation in soils: it is the balance between the availability of the element in the decomposing substrate (measured by its element/carbon ratio) and the microbial requirements for it that determines if mineralisation or immobilisation predominates at a given time.[2] Perhaps this assimilation (also called carbon-element) theory, to which we have already referred in Chapter 3, could explain sufficiently well the dynamics of N, P, S if the soil organic matter were an ideal, homogeneous substrate. Numerous investigations, however, show results that cannot be explained by, and even contradict, the old theory. The soil organic matter is a very complex substrate consisting of a mixture of qualities and cohorts characterised at a given time by different mineralisation-immobilisation statuses. It is not surprising then, that the mineralisation of C, N, P, S in soils of apparently the same characteristics may react differently in response to the same kind of perturbation.

The organic matter of the soil has a large potential to retain inorganic nitrogen (up to 30 to 70% of added fertiliser)[3] by means of both biotic (microbially mediated) and abiotic (chemical) reactions.[4] Since the soils are energy limited, this process is rarely accompanied by any increase in microbial activity.[5] In spite of this, it has been found that much of the newly added ^{15}N occurs in soils as insoluble components of microbial tissues.[6] If nitrogen is available, fungi are capable of increasing their nitrogen concentration beyond the needs imposed by their carbon status.[7] A consequence of this, particularly in forest ecosystems, is a large reduction in the availability of supplied N-fertilisers to vegetation.[8]

Liming initially increases C mineralisation in soils; after a certain period of time the effect fades out and the mineralisation rate may eventually become lower than the same rate in the untreated soil.[9] To explain this initial stimulation it has been hypothesized that liming increases the availability of C sources in the soil to the micro-organisms.[10] The reported effects of liming upon the other elements are more contradictory; some authors have found that liming increases net N mineralisation[11] whereas others have found decreases.[12] Liming can affect the organic matter in different ways. One possibility is that lime changes the physical accessibility of different compounds. If this is the case, the quality of the organic matter remains unchanged but the microbes can grow more rapidly because more of the carbon will be directly available. Such effects should be accounted for by changing the function $u(q)$. Another possibility is that lime is changing the chemical properties of the organic matter; conformational changes of molecules would be a typical example. In this case, we should describe the effect of lime as changing the carbon distribution, $\rho_C(q,t)$. At the moment, we have no information telling us which of these alternatives is the most important one. However, changes in $u(q)$ are a much simpler problem to analyse.

7.2   Mineralisation-immobilisation in single litter cohorts

Equation (4.37) provides us with a definition of net mineralisation, m. Let us repeat this equation in a slightly modified notation

$$m(\hat{q}) = \frac{f_C u(\hat{q})}{e(\hat{q})} N(t) - f_n u(\hat{q}) C(t) \qquad (7.1)$$

The element n being incorporated in microbial biomass at a rate $f_n u C$ can come from organic forms in the substrate or from the inorganic pool in the soil solution. Jansson (1958) and others observed that considerable quantities of labelled N are immobilised when soil is incubated with labelled inorganic N even if there is net mineralisation of N by the soil. We interpret this as an indication that the microbes prefer inorganic forms of the element; in the extreme all nitrogen is obtained from inorganic sources. We will adopt this extreme view to see how far it will go in explaining observations and from now on assume that the second term in the right hand side of (7.1) is the rate of immobilisation of inorganic N.

Support for such an interpretation of (7.1) can be obtained from a study by Berg (1988) of the dynamics of nitrogen in ^{15}N labelled Scots pine needles given an unlimited supply of inorganic ^{14}N. The evolution of the isotope distribution permits us to distinguish the dynamics of the nitrogen that is initially present in the needles from that of the immobilised nitrogen. Let $^{15}N(\hat{q})$ denote the nitrogen that was present in the needles at the beginning of the decomposition (note that as time goes on, decomposers recycle this initial nitrogen many times) and let $^{14}N(\hat{q})$ represent the amount of immobilised nitrogen, i.e. the nitrogen in the needles that was not present in the needles at $\hat{q} = q_0$. Because of the mineralisation-immobilisation cycle, a certain fraction x of the inorganic pool will come to consist of the nitrogen that was initially in the needles. Equation (7.1) can then be used to describe the dynamics of ^{14}N and ^{15}N if the immobilisation term is changed to $f_N u C x$ in the equation for ^{15}N and to $f_N u C(1 - x)$ in the equation for ^{14}N. This gives

$$\frac{d^{15}N}{dt} = \frac{f_C}{e(\hat{q})} u(\hat{q})^{15}N(t) - x f_N u(\hat{q}) C(t) \qquad (7.2)$$

$$\frac{d^{14}N}{dt} = \frac{f_C}{e(\hat{q})} u(\hat{q})^{14}N(t) - (1 - x) f_N u(\hat{q}) C(t) \qquad (7.3)$$

These equations are non-linear and difficult to solve. Since the initial amount of nitrogen in the litter cohort is small compared with the initial amount in the inorganic pool, the inorganic pool can only contain small amounts of nitrogen that initially were in the litter, and x remains close to zero all the time. Making the approximation $x = 0$, the equations can be solved; with Model II for decomposers we get

$$^{15}N(\hat{q}) = r_0 \left[\frac{\hat{q}}{q_0} \right]^{1/\eta_{11}} C(\hat{q}) \tag{7.4}$$

$$^{14}N(\hat{q}) = \frac{f_N}{f_C} \left[1 - \left(\frac{\hat{q}}{q_0} \right)^{1/\eta_{11}} \right] C(\hat{q}) \tag{7.5}$$

Note that by adding (7.4) and (7.5) we get (4.51), which gives the dynamics of the total N in the litter cohort.

In Figure 7.1, the theoretically calculated and empirically measured amounts of nitrogen (^{14}N, ^{15}N) are shown as a function of relative mass loss $(1 - C/C_0)$. The figure displays the phase of nitrogen immobilisation and the subsequent phase of nitrogen mineralisation. The theoretical values of immobilisation are larger than the experimental values. This is, of course, what one expects with the assumption that all the nitrogen incorporated in the biomass is taken from the inorganic sources. However, as an approximation, it seems satisfactory.

Consider now the inverse problem, where an unlabelled litter cohort decomposes in the presence of a pool of ^{15}N labelled inorganic nitrogen. Let us refer to this as the experimental system and compare it with an equal litter cohort decomposing in a pool of unlabelled nitrogen (the control). The amounts of ^{14}N in the inorganic pools at $\hat{q} = q_0$ are the same in the control and the experiment. As decomposition goes on, the amount of inorganic ^{14}N becomes larger in the experiment than in the control. Jenkinson et al. (1985) named this pool substitution effect "added nitrogen interaction" (ANI and for which we will use the symbol A) and defined it as the difference between the amounts of inorganic ^{14}N in the control and the experiment, Figure 7.2. They calculated A in an approximate way, valid only for short time intervals. By making use of our equations we can calculate A over the whole span of decomposition. Since, with respect to nitrogen, we are dealing with a closed system and the initial amounts of inorganic ^{14}N are the same, the following relationship holds

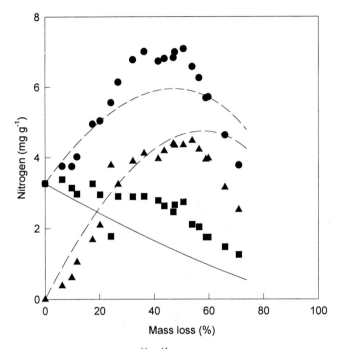

Figure 7.1 Development of the total (^{14}N+^{15}N; ● and broken line) and of the initially present nitrogen (^{15}N; ■ and solid line) in needle litter; the difference (^{14}N; ▲ and broken line) is a measure of the immobilisation process (data from Berg 1988). The theoretical curves are calculated from equations (4.45), (7.4) and (7.5), respectively. The parameters in these equations have the following values: $f_C = 0.5$, $f_N = 0.00225$, other parameters from Table 5.1 (from Bosatta & Ågren 1995b).

$$A = N_c(\hat{q}) - N_e(\hat{q}) \tag{7.6}$$

where N_c, control, is given by (4.51) and N_e is the evolution over time of the organic ^{14}N in the experiment. To calculate N_e let us make the following simplifying assumption: the experiment is prepared in such a way that the amount of organic nitrogen is small compared with the amount of inorganic nitrogen. Then, the concentration of ^{15}N in the inorganic phase can be considered as constant and equal to the initial value, x_0, over the whole period of decomposition and N_e becomes similar to N_c but with f_N substituted with $(1 - x_0)f_N$. We get

$$A = x_0 \frac{f_N}{f_C} \left[1 - \left(\frac{\hat{q}}{q_0} \right)^{1/\eta_{11}} \right] C(\hat{q}) \tag{7.7}$$

Note that A follows the same behaviour as the immobilised nitrogen in Berg's experiments (equation 7.5 and Figure 7.1). Jenkinson et al. (1985) observed that A should increase with the fraction of labelled inorganic nitrogen and with any factor increasing the rate of microbial immobilisation. This is precisely what (7.7) predicts because litter fractions of higher q_0 will produce higher immobilisation rates.

We have thus shown that the term $f_N u(\hat{q})C$ in (7.1) can be identified with the rate of immobilisation. We will now consider how to incorporate the availability of inorganic nitrogen in this immobilisation term.

Problem 7.1
Derive the equations for $d^{15}N/dt$ and $d^{14}N/dt$ and verify that in the limit of $x = 0$ the solutions are as given by (7.4) and (7.5).

Problem 7.2
Derive equations (7.6) and (7.7).

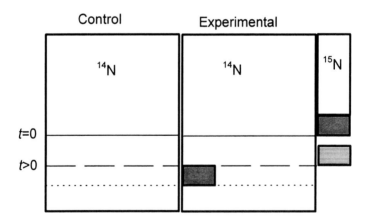

Figure 7.2 Illustration of the "added nitrogen effect". At $t = 0$ the nitrogen is divided between inorganic (above the solid line) and organic (below the solid line) N. At a later time ($t > 0$) organic N has been converted to inorganic N and the partitioning has moved to the broken line. However, there has also been a gross immobilisation (the area between the broken and dotted line), part of which is ^{15}N (the hatched area). This immobilisation of ^{15}N corresponds to an equal increase in ^{14}N in the inorganic form.

7.3   N retention in the soil organic matter

The exchange of N between soil organic matter and the inorganic nitrogen occurs by two different processes. Firstly, the microbes assimilate and reduce the availability of inorganic nitrogen. To include effects of the size of the inorganic pool we multiply the immobilisation rate of (7.1) by a function $F(N_i)$ of the inorganic pool N_i ($g_N g_{soil}^{-1}$). This function has the general properties of being zero when N_i is zero and asymptotically approaching 1 when N_i becomes very large. We view this as an adjustment of the microbial nitrogen concentration.

Secondly, there are purely abiotic reactions between soil organic matter and inorganic N. Since the compounds giving rise to the abiotic fixation are partly produced during decomposition, the dependence on quality of the rate of abiotic immobilisation i_a ($g_N g_{soil}^{-1} d^{-1}$) may be important. We define:

$$i_a = F(N_i)\int s(q)\rho_C(q,t)dq \cong s(\hat{q})F(N_i)C(t) \qquad (7.8)$$

where $s(\hat{q})$ ($g_N g_C^{-1} d^{-1}$) is the specific rate of abiotic fixation, and, for simplicity, we have assumed that i_a is controlled by N_i in the same way as microbial immobilisation.

Using (7.8) and (7.1) we get the dynamic equation for the inorganic pool of nitrogen:

$$\frac{dN_i}{dt} = \frac{f_C}{e(\hat{q})}u(\hat{q})N - \left[s(\hat{q}) + f_N u(\hat{q})\right]F(N_i)C \qquad (7.9)$$

Suppose that, at time t an amount Δ_T of N is added to the inorganic pool; the added N becomes distributed between the organic, $N + \Delta_o$, and the inorganic pool, $N_i + \Delta_i$. Since (7.9) describes a closed system (uptake and leaching have been neglected) $\Delta_T = \Delta_o + \Delta_i$. The turnover of the inorganic pool is rapid compared with the organic components so one may assume that it relaxes almost immediately to the value imposed by the surroundings. We then set dN_i/dt to zero, keep only first order terms in Δ_o and $\Delta_{i,}$ and assume that before the perturbation N_i was at equilibrium to get

$$\frac{\partial F}{\partial N_i}\left[s(\hat{q}) + f_N u(\hat{q})\right]C\Delta_i = \frac{f_C}{e(\hat{q})}u(\hat{q})\Delta_o \qquad (7.10)$$

Assume now that F is of the form of the Michaelis-Menten equation,[13] i.e.

$$F = \frac{N_i}{K + N_i} \tag{7.11}$$

where K is the affinity constant. Then,

$$\frac{\Delta_o}{\Delta_i} = \frac{[s(\hat{q}) + f_N u(\hat{q})]}{\frac{f_C}{e(\hat{q})} u(\hat{q})} \frac{KN_i}{(K + N_i)^2} \frac{C}{N_i} \tag{7.12}$$

Using this equation we can calculate the fraction Δ_o/Δ_T of added N that the soil retains in terms of different properties of the organic matter (C, N_i and $\hat{q}$) at the time of addition.

We must now calculate the parameters in (7.12). We estimate the $s(\hat{q})$ function from Axelsson & Berg (1988) who, in short-term (24 hours) experiments, studied the abiotic fixation (samples were treated with $HgCl_2$ before addition of N) of ^{15}N in needle litter at different stages of decomposition.

We assume that $s(\hat{q})$ has the following form:

$$s(\hat{q}) = s_0 \hat{q}^\gamma \tag{7.13}$$

where s_0 is the rate at a reference quality ($\hat{q} = 1$) and γ defines the shape of the absorption function. Elimination of $\hat{q}$ between (4.51) and (7.13) provides the theoretical relation $s(r)$ between the specific rate and the N:C ratio. Figure 7.3 shows $s(r)$ calculated in this way together with the experimental data of Axelsson & Berg (1988). The values of the parameters are the same as in Figure 7.1, except that r_0 has the values measured for undecomposed needles, 7.8 mg g^{-1}, $s_0 = 40$ $\mu g_{NgC}^{-1}d^{-1}$ that has been adjusted to give a good fit, and $\gamma = 7$. The value of s_0 is similar to what Schimel & Firestone (1989) estimated as a maximum rate of abiotic fixation (~ 3 $\mu g_{NgC}^{-1}h^{-1}$) in samples from the organic horizon of a forest soil.

We can now estimate N retention in soils near steady state. Under such conditions microbial growth is mostly limited by energy sources. We have defined microbial growth rate as $u_0 \hat{q}^\beta$ and estimated u_0, the rate at a reference quality of 1, to $5 \cdot 10^{-4}$ g $g_C^{-1}d^{-1}$ (Model II, Table 5.1) under field conditions (5 °C). If we set f_N in (7.9) to 0.025, then the maximum rate of microbial immobilisation is 13 $\mu g_{NgC}^{-1}d^{-1}$ which, considering a Q_{10} factor ≥ 2, compares with the value ~ 200 $\mu g_{NgC}^{-1}d^{-1}$ determined by Schimel & Firestone (1989) under laboratory conditions (20 °C).

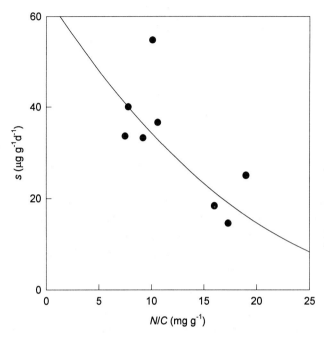

Figure 7.3 The specific rate of abiotic nitrogen fixation, s, in needle litter as a function of the N:C ratio of the needles (data ● from Axelsson & Berg 1988). The theoretical curve (line) is constructed from equations (4.45) and (7.13) and with parameter values as in Figure 7.1 (from Bosatta & Ågren 1995b).

It is known from experiments that soils differ greatly in their capacity to retain added inorganic nitrogen, such that forest soils (F) tend to retain more nitrogen than agricultural (A) soils.[14] We want to compare the retention capacity of A and F soils in terms of equation (7.12). Since $\beta = \gamma = 7$ and $e(\hat{q}) = e_0$, the first factor on the right hand side of (7.12) is independent of $\hat{q}$. If we assume that the remaining variables are the same in both types of soils then the first factor in (7.12) is of the same order of magnitude in both systems. The affinity constant K was estimated at 0.4 mM by Schimel & Firestone (1989). In A soils N_i can be > 1 mM,[15] which means that the second factor in (7.11), $KN_i/(K + N_i)^2$ can vary between 0.04 and 0.4. Normally, in F soils $N_i \ll K$ and $KN_i/(K + N_i)^2$ can be estimated to 0.1 or smaller. So, again, the second factor of (7.12) is of the same order of magnitude in both systems.

The difference in retention capacity is due to the last factor in (7.12), C/N_i, which can be more than one order of magnitude larger in F soils. Why? Setting (7.9) to zero we get

$$\frac{C}{N_i} = \frac{N}{N_i}\frac{C}{N} = \frac{e_0}{f_C}\frac{s(\hat{q}) + f_C u(\hat{q})}{u(\hat{q})}\frac{1}{K + N_i}\frac{C}{N}C \qquad (7.14)$$

In (7.14) all the terms are of the same magnitude for A and F soils, except the last, C, which varies as $q_0^{-\beta}$ (4.53). Hence,

$$\frac{(C/N_i)_F}{(C/N_i)_A} \propto \left(\frac{q_{0A}}{q_{0F}}\right)^{\beta} \qquad (7.15)$$

Considering the low quality of the woody components in forest ecosystems, it is not unreasonable to expect $q_{0A}/q_{0F} \cong 1.5$, and since $\beta = 7$ the $C:N_i$ ratio will be at least one order of magnitude larger in F-soils.

Figure 7.4 shows the fraction of N-fertiliser retained in the soil organic matter in terms of the $C:N_i$ ratio together with data from some agricultural and forest systems (data from Melin 1986 are not the directly measured fractions but the fractions corrected for vegetation uptake and other sinks, i.e. the fractions calculated by using the measured values of Δ_o and Δ_i). The figure clearly displays the effect of the difference in C/N_i upon the retention capacity of both soils and the relatively minor effect of K.

7.4   Perturbation in carbon accessibility

Since an increase in $u(q)$ basically means an increase in decomposer activity, at least initially, the carbon mineralisation rate m_C must also increase. In a soil sample without further inputs of organic matter this increased activity will, relative to the unperturbed situation, move the sample more rapidly down towards lower qualities and, hence, lower rates. It is, therefore, not difficult to explain the observations concerning the evolution of carbon mineralisation referred to in the beginning of this chapter. It is more difficult to explain the immobilisation phase sometimes observed for the other elements. However, if we assume that an increase in $u(q)$ is not uniform but affects the highest qualities mostly, it is possible to create such a scenario. Liming is one perturbation that is supposed to preferentially make the more easily degradable substrates available. Let us therefore apply this reasoning to some experiments with liming.

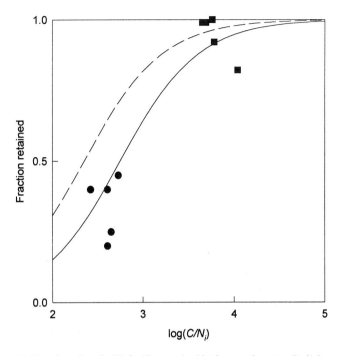

Figure 7.4 Fraction of applied N-fertiliser retained in the organic matter (Δ_o/Δ_T) as a function of the logarithm of the C:N_i ratio of the soil. Data for forest systems (■) from: Melin (1986) and Aber et al. (1993); data for agricultural soils (●) from Jansson (1958). The curves are theoretical values calculated from equation (7.12) (forests: broken line and $KN_i/(K + N_i)^{-2} = 0.1$; agricultural soils: solid line and $KN_i/(K + N_i)^{-2} = 0.04$; other parameters as in Figure 7.1) (from Bosatta & Ågren 1995b).

We assume that the effect of the perturbation (the liming) is to increase u_0 of (4.37) to a higher value u_p ($u_p > u_0$), increasing in this way the decomposer activity. However, we let only the highest qualities be affected. When the quality of a litter cohort reaches a limit value q_t ($0 < q_t < q_0$) the effects of liming cease and u_p is replaced by u_0. We will moreover assume that the samples we are considering are cut off from further litter inputs at the same time as the liming is performed.

At time t after the liming there will then be three kinds of cohorts in the sample: i) cohorts old enough (older than t_l) not to be affected by the perturbation, i.e. cohorts for which $q < q_t$ already at $t = 0$; ii) cohorts that were initially affected but, at time t_l have passed beyond q_t; and iii) young cohorts which, at time t, are still affected $(q > q_t)$; this latter kind of cohort will eventually disappear, Figure 7.5. We can now calculate the rates of mineralisation of treated and untreated organic matter.

We have from (4.47)

$$C(t) = \int_0^{t_p} I(t-a)g(q_{a,t})da \qquad (7.16)$$

where t_p is the age of the oldest litter cohort and where, on q, we indicate both the age, a, of the cohort at the time of the perturbation and the time that has passed since the start of the perturbation, t. As usual we will assume a constant rate of litter input, I_0, but we can not evaluate (7.16) directly as was done in (4.53) but have to split the integral into three parts according to the ages of the cohorts as discussed above and in Figure 7.5. We can then use the following reformulation to calculate the mineralisation rate

$$m_C(t) = -\frac{dC}{dt} = I_0 \int_0^{t_p} \frac{dg(q_{a,t})}{dt} da = I_0 \int_0^{t_p} \frac{dq/dt}{dq/da} dg \qquad (7.17)$$

To proceed, we need to find dq/dt and dq/da for the three regions. This is easy for the cohorts belonging to region I because they have always evolved at rate u_0 and the relation between q, a, and t is given directly by (4.52)

$$q_i(a,t) = \frac{q_0}{\left[1+\beta f_C \eta_{11} q_0^\beta u_0(a+t)\right]^{1/\beta}} \qquad (7.18)$$

Similarly, for cohorts in region III, the expression is almost identical except that these cohorts have evolved at rate u_0 up to the perturbation and at rate u_p afterwards

$$q_{iii}(a,t) = \frac{q_0}{\left[1+\beta f_C \eta_{11} q_0^\beta (u_0 a + u_p t)\right]^{1/\beta}} \qquad (7.19)$$

The expression becomes somewhat more complicated for cohorts in region II because, before the perturbation, these evolved at rate u_0, then for some time t_1 at rate u_p until they hit the border q_t, after which they evolve again at rate u_0 for the remaining time t_2 $(t = t_1+t_2)$. We then have

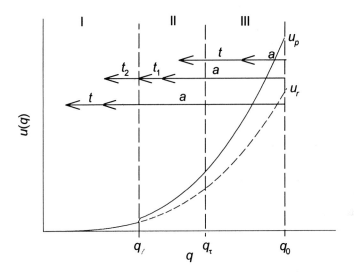

Figure 7.5 Description of the three regions of qualities to consider when analysing the effects of a perturbation of $u(q)$. See text for further description.

$$q_{ii}(a,t) = \frac{q_0}{\left[1 + \beta f_C \eta_{11} q_0^{\beta} (u_0 a + u_p t_1 + u_0 t_2)\right]^{1/\beta}} \qquad (7.20)$$

We can calculate t_1 (and t_2) from the condition that at time t_1 $q_{ii}(a,t_1) = q_l$, thus

$$\beta f_C \eta_{11} q_0^{\beta} (u_0 a + u_p t_1) = \left(\frac{q_0}{q_\ell}\right)^{\beta} - 1 \qquad (7.21)$$

Combining (7.20) and (7.21) finally gives

$$q_{ii}(a,t) = \frac{q_0}{\left[\left(\frac{q_0}{q_\ell}\right)^{\beta}\left(1 - \frac{u_0}{u_p}\right) + \frac{u_0}{u_p} + \beta f_C \eta_{11} q_0^{\beta}\left(\frac{u_0^2}{u_p} a + u_0 t\right)\right]^{1/\beta}} \qquad (7.22)$$

We can now evaluate (7.17) to get

$$\frac{m_C^p(t)}{I_0} = \begin{cases} -g(q_i(t_p,t)) + g(q_i(t_\ell,t)) - \dfrac{u_p}{u_0}\left[g(q_{ii}(t_\ell,t)) - g(q_{iii}(0,t))\right] & \text{for } t \le t_l \\[3mm] -g(q_i(t_p,t)) + g(q_i(t_\ell,t)) - \dfrac{u_p}{u_0}\left[g(q_{ii}(t_\ell,t)) - g(q_{ii}(0,t))\right] & \text{for } t > t_l \end{cases}$$

$$(7.23)$$

and where the superscript p indicates the perturbed sample. The mineralisation rate for an untreated reference sample is, of course, also given by (7.23), but with $u_p = u_0$. Since we have nowhere used the specific properties of g in (7.23), we just need to replace g with h_n to get the corresponding expression for the mineralisation of an element.

All parameters entering (7.23) are known with the exception of u_p and q_l. The value of q_l affects mainly the length of the period that a substrate spends at the higher decomposer growth rate, while u_p affects the magnitude of the initial increase in respiration rate. Persson et al. (1991) followed the evolution of C and N mineralisation in limed needle litter incubated in the laboratory at 15 °C; we use their data on C mineralisation in order to estimate u_p and q_l (Figures 7.6 and 7.7). Other parameters in this figure are as in Figure 7.3 with the exception of u_0. Considering a Q_{10} factor between 2 and 3 we can assume $u_0 q_0^\beta$ to be approximately 3 times larger (0.002 $g \cdot g^{-1} \cdot d^{-1}$) than the field value from Table 5.1.

The parameter values that we have chosen to get a good fit for these two figures show that it is only over a small quality interval that liming is effective, but that where it changes the carbon accessibility the effect is very large ($u_p = 3u_0$). Because of the crude way we have introduced the effect of liming, we should be a bit cautious when emphasising the quantitative effects, although they provide some hints that systems dominated by low quality substrates may react less to liming than those composed of high quality substrates.

Since q_l is very close to q_0, the immediate response of the nitrogen-rich system to liming at $t = 0$ is almost nitrogen immobilisation, a fact that probably escapes experimental observation. Later, the level of mineralisation is larger in the N-rich material. As was discussed at length in 7.3, soils rich in N should have a higher value of f_N than poor soils. The N_{ss}/C_{ss} values for the rich and poor soils are 0.04 and 0.016, respectively (Persson et al. 1991); using (4.56) we should have $f_N/f_C \cong 0.05$ in the N-rich stand and $f_N/f_C \cong 0.013$

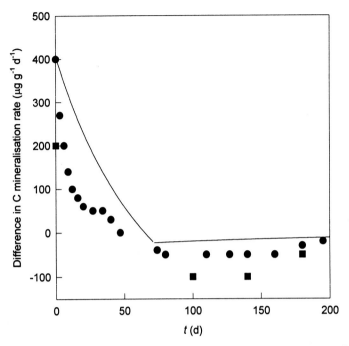

Figure 7.6 Difference in C mineralisation rate between limed and untreated needle litter sampled from N-rich (●) and N-poor (■) soils (Data from Persson et al. 1991). The theoretical curve is calculated with equation (7.23). Parameter values are $u_0 q_0^\beta = 0.002$ g·g^{-1}·d^{-1} (level of decomposer growth rate in untreated samples), $u_p q_0^\beta = 0.006$ g·gC^{-1}·d^{-1} (in limed samples) and $q_l = 0.94 q_0$. Other parameters as in Figure 7.3.

in the poor stand, which is approximately half the values used in Figure 7.7. We would, however, again like to stress the qualitative nature of this analysis.

The above analyses can be applied to other perturbations than liming. For example, acidification is interpreted as producing a decrease in $u(q)$, $u_p < u_0$. From (7.23), $\Delta m_C(0) < 0$, which means that the respiration rate is decreased. On the other hand, $\Delta m_N(0) > 0$ (if $r_{0N} < h_N(q_l)$) and the immediate response is an increase in nitrogen mineralisation instead of an increase in immobilisation, as was the case for liming.

Any perturbation affecting $u(q)$ is going to produce effects of the kind described above for liming. This may be the case for certain kinds of fertilisations. For example, White et al. (1988) observed oscillations of the

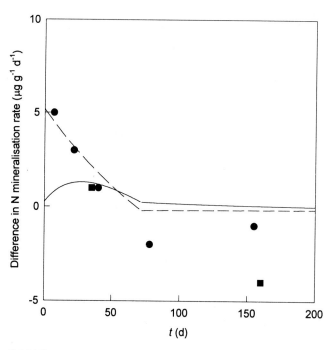

Figure 7.7 Difference in net N-mineralisation rate between limed and untreated needle litter sampled from N-rich (●) and N-poor (■) soils and incubated in the laboratory (Persson et al. 1991). The theoretical curves, N-rich stand (solid line) and N-poor stand (broken line), are calculated with equation (7.23). Parameter values are $r_{0N} = 0.02$ (initial N:C ratio of a litter cohort), $f_N/f_C = 0.1$ in the rich stand, $f_N/f_C = 0.05$ in the poor stand. Other parameters as in Figure 7.6.

kind shown in Figures 7.6 and 7.7 after they had perturbed the field with additions of C and N.

Problem 7.3

Consider a perturbation that changes u_0 to u_p ($u_p > u_0$) in (4.37) in a soil sample at steady state and where input is disrupted at the time of the perturbation. Calculate the time it takes until the carbon mineralisation rate of the unperturbed system exceeds that of the perturbed one.

Problem 7.4

Verify and derive the explicit expressions for carbon and nitrogen mineralisation in (7.23).

Problem 7.5

Derive the explicit conditions for a net nitrogen immobilisation to occur at $t = 0$ with the kind of perturbation described in this section.

Problem 7.6

Calculate for a soil at steady state the ratio between net nitrogen mineralisation and gross nitrogen mineralisation.

Problem 7.7

Show that for small applications of fertilisers, the time evolution of the remaining nitrogen, $\Delta_i(t)$, is given by the equation

$$\Delta_i(t) \approx \Delta_i(t_0) \left[\frac{\hat{q}(t_0 + t)}{\hat{q}(t)} \right]^{1/\eta_{11}e_0}$$

where t_0 is the time when the fertiliser is applied and $\Delta_i(t_0)$ is given by (7.11). Hint: Assume that $\Delta_i(t) << \Delta_i(t) + N_i(t)$.

7.5   Variable decomposer growth rate

In both Chapters 6 and 7 we have considered perturbations where the basic decomposer growth rate, u_0, has been affected. These cases have been treated on their own merits. We can, however, discuss them as special cases of a more general problem. Suppose that we can write the decomposer growth rate per unit carbon, u, in the following form

$$u(q, t) = u_1(q)u_2(t) \tag{7.24}$$

We can then evaluate (4.27)

$$\int_0^t u_2(\tau)d\tau = \int_{q_t}^{q_0} \frac{dq}{f_C \eta_1(q)u_1(q)} \tag{7.25}$$

The integral on the right-hand side has not changed from previous applications and the same decomposer models can be used. The integral on the left-hand side is, however, a new acquaintance. In all previous applications we have assumed that $u_2(t)$ is constant, or piecewise constant in some of the applications in Chapters 6 and 7. From (7.25) we can, however, see that more complicated situations can be handled; any variation in decomposer growth rate can, of course, be handled through numerical integration of (7.25). More interesting are those cases where we can get closed expressions for the left-hand side of (7.25). One such case is when the decomposer growth rate varies sinusoidally (in response either to diurnal or to annual climatic cycles). We can then evaluate the integral[16]

$$\int_0^t (T_m + T_a \sin \omega\tau)\,d\tau = T_m t + \frac{1}{\omega} T_a (1 - \cos \omega t) \qquad (7.26)$$

which is particularly simple if we only are interested in the results over whole cycles ($\omega t = n2\pi$). In such cases, we can also evaluate (7.25) when there is an exponential response to a sinusoidally varying factor[17]

$$\int_0^{n2\pi/\omega} e^{\alpha(T_m + T_a \sin \omega\tau)}\,d\tau = \frac{n2\pi}{\omega} e^{\alpha T_m} I_0(\alpha T_a) \qquad (7.27)$$

where I_0 is a zero order Bessel function.[18]

NOTES

[1] Doryland (1916), cited by Jansson (1958).

[2] See also Swift et al. (1979) and Parnas (1975).

[3] Carter et al. (1967), Reddy & Patrick (1978), Melin & Nömmik (1988), Stevenson (1982).

[4] Nömmik & Nilsson (1963), Stevenson (1982), Schimel & Firestone (1989), Johnson (1992). The importance of biotic processes (microbial mineralisation-immobilisation) was extensively analysed by Jansson (1958) in his classical ^{15}N experiments. Abiotic incorporation has been explained as condensation reactions occurring between phenols and derivatives, originating from partially degraded lignin and fungal products, with either amino acids or ammonia (Mortland & Wolcott 1965, Burge & Broadbent 1961, Kelley & Stevenson 1987).

[5] Fog (1988).

[6] He et al. (1988).

[7] Levi & Cowling (1969), see also Paustian & Schnürer (1987). Saggar et al. (1981) observed a similar effect of increased microbial sulphur concentration following additions of sulphate to the soil.

[8] Tamm (1963), Ingestad et al. (1981), Aber et al. (1993). The capacity to retain inorganic nitrogen is also critical in determining when an ecosystem becomes nitrogen saturated (Ågren & Bosatta 1988).

[9] Nömmik (1978), Lohm et al. (1984).

[10] Persson & Wirén (1989).

[11] Nyborg & Hoyt (1978). Increased immobilisation of S has also been reported in limed soils (Nömmik et al. 1984, Nilsson pers. comm.).

[12] Nömmik (1978).

[13] Schimel & Firestone (1989).

[14] Melin et al. (1983), Melin & Nömmik (1988), Nömmik & Larsson (1989), Jansson (1958).

[15] Jansson (1958).

[16] Ågren (1989).

[17] Ågren & Axelsson (1980).

[18] e.g. Abramowitz & Stegun (1972).

PART III

THE PLANT

"I can't help it," said Alice very meekly:
"I'm growing."
"You have no right to grow here,"
 said the Dormouse.
"Don't talk nonsense," said Alice more boldly:
"you know you're growing too."
"Yes, but I grow at a reasonable pace,"
 said the Dormouse:
"not in that ridiculous fashion."

(Lewis Carroll *Through the looking glass*)

8

Theory for plant growth

The importance of nutrient availability for plant growth has been recognised and studied in a systematic way at least since the middle of the last century when concepts such as Liebig's law of the minimum became established. The quantitative relations were, however, expressed at the practical level; such as the relationship between added amount of fertiliser and yield, e.g. the Mitscherlich curve.[1] When more mechanistic explanations in the form of mathematical models started to appear in the 1960s, nutrition was generally left out, in many cases it was explicitly assumed that crop plants were growing at optimum nutrition.[2] Interest was instead focussed on the carbon balance of the plants and efforts were directed towards calculations of photosynthesis and light distributions within canopies.

Another characteristic feature of the early models of plant growth was their design, that made them solvable only through computer simulations. Numerous books and journal papers from the IBP period witness that such an engineering approach was seen as the future for theoretical studies of plant growth. This trend was, however, not unchallenged and arguments for models that permitted analytical investigations were raised.[3]

We pursue in this chapter the analytical approach, with emphasis on nutrient control of plant growth, and develop the concept of nitrogen productivity, which is our key in the analysis of plant growth. A comparison with a model based on the nitrogen control of the plant's carbon balance shows why the nitrogen productivity works. An objective is to develop an understanding of the concept and to show how to use it by exploring qualitative examples. We also extend the nitrogen productivity concept to situations where other mineral nutrients might be limiting and to situations of interactions with other environmental factors. Throughout this chapter the plant's access to nitrogen is assumed to be outside the control of the plant but determined by some external constraints. Plants are also assumed to be small, such that interactions within and between individuals can be neglected.

8.1   Nitrogen productivity

Before defining the concept of nitrogen productivity, it is worthwhile exploring the alternatives that are available to us when expressing the influence of nutrition on plant performance. We will use plant growth rate (or change in plant biomass per unit of time) as our indicator of plant performance. A simple evaluation of possible variables for describing nutrient status is dimensional analysis.[4] Table 8.1 shows the units of the variables required to link plant growth rate with the solute concentration in the root medium, the amount of nitrogen in the plant, or the plant nitrogen concentration. Of these variables, it seems that only the variable associated with plant nitrogen amount has units that can open avenues for further analysis. This is the variable that we will denote *nitrogen productivity*.

Let us now see how the concept of nitrogen productivity can connect the plant's carbon and nitrogen cycles. Our basic assumption is that the fixation of carbon in the plant is controlled by the amount of proteins present. As long as the plant grows under reasonably stable conditions evolutionary optimisation should have created a balance between the different types of proteins (enzymes) required, whereby, in principle, any of the proteins in the plant could serve as a model from which to calculate plant performance (cf. Section 2.2). With nitrogen as the characteristic element of proteins, it is natural to let the rate of formation of new biomass be related to the amount of nitrogen

Table 8.1 *Units of variables used to describe plant nutrition and functions connecting plant nutrition to plant growth rate (from Ingestad & Ågren 1992).*

Control variable		Connecting function	
Variable	Dimension	Dimension	Variable
Solute concentration	$\dfrac{[mol]}{[volume]}$	$\dfrac{[mass][volume]}{[mol][time]}$	—
Plant nitrogen concentration	$\dfrac{[nitrogen]}{[mass]}$	$\dfrac{[mass]^2}{[nitrogen][time]}$	—
Plant nitrogen amount	$[nitrogen]$	$\dfrac{[mass]}{[nitrogen][time]}$	Nitrogen productivity

present in the plant. Our intention is also to operate at a high level of integration and to avoid having to look at a large number of detailed processes. The formalisation of these ideas is the concept of nitrogen productivity, which we define as the amount of biomass produced per amountof nitrogen in the plant per unit of time. However, all the nitrogen in a plant is not active in growth. We account for this by assuming that a certain minimum concentration of nitrogen in the plant, $c_{N,min}$, should be discounted. Formally, we then express the relation between growth and nitrogen as

$$\frac{dW}{dt} = P_N (N - c_{N,min}W) \tag{8.1}$$

where W is plant biomass, N amount of nitrogen in the plant, t time, and P_N the nitrogen productivity. At internal nitrogen concentrations below $c_{N,min}$, the relative growth rate is 0.

Equation (8.1) is our basic equation for the analysis of relationships between plant nutrition and plant growth. We will start with some additional, simplifying restrictions in order to retain attention on the concept itself; some of these assumptions will be relaxed later. The plant is therefore assumed to suffer from no mortality. Limitation from other factors such as light, temperature, moisture, etc., may be at hand, but they do not vary over the time period in which we are interested. Variations in growth rate due to seasonality are also outside the scope here. Other mineral nutrients are assumed to be in sufficient supply; we will return to this point later in this chapter. However, we leave open the possibility that nitrogen productivity may change due to the growth of the plant or the uptake of nitrogen.

The potential use of (8.1) depends on the degree of simplicity of the models we can use for P_N and for the uptake rate of nitrogen. Obviously, the simplest model for P_N is a constant. Let us, therefore, explore that model first. Hence,

$$P_N = \text{constant} \tag{8.2}$$

The plant is thus characterised by two parameters, P_N and $c_{N,min}$, which under constant environmental conditions are constant but, of course, vary between species.

By combining (8.1) and (8.2) we can now derive an expression for the relative growth rate, R_W, of the plant. The relative growth rate expresses relations in terms of intensive properties only and is a more convenient basis for much of the discussion to follow; we do not need to worry about the absolute plant size.

$$R_W = \frac{1}{W}\frac{dW}{dt} = P_N\left(\frac{N}{W} - c_{N,min}\right) = P_N(c_N - c_{N,min}) \qquad (8.3)$$

Hence, the relative growth rate is a linear function of the internal nitrogen concentration, c_N, with nitrogen productivity as the slope and $c_{N,min}$ as the intercept of the abscissa. Equation (8.3) cannot be valid for all nitrogen concentrations. In Figure 8.1 we illustrate graphically the range of validity we attribute to (8.3) and what happens outside that range.[5] At internal nitrogen concentrations below $c_{N,min}$, nitrogen is clearly in deficiency and there is no response to increased nitrogen concentrations. At internal nitrogen concentrations above $c_{N,min}$, the plant will respond by increasing its relative growth rate linearly with the internal nitrogen concentrations up to some maximum relative growth rate, r_m. This relative growth rate is attained at a nitrogen concentration that we call the optimal one, $c_{N,opt}$; the optimal nitrogen concentration is the minimum concentration consistent with the maximum relative growth rate. At still higher internal nitrogen concentrations, the plant will no longer increase its growth rate. In this range of concentration, nitrogen is in sufficiency. This is sometimes called luxury consumption, but we prefer the more neutral term sufficiency, as luxury has a connotation of something not

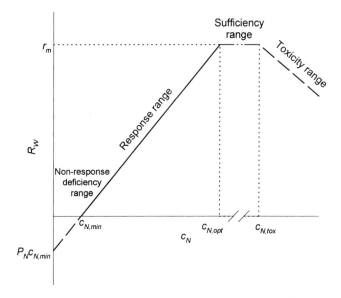

Figure 8.1 The principal relationship between relative growth rate and plant nitrogen concentration with different response ranges indicated.

needed. Since improved growth conditions in other respects than nitrogen nutrition or a decrease in the rate of nitrogen uptake can move the plant down into the response region again, the nitrogen acquired in excess at one moment can be utilised later and may only temporarily be an excess. Many plant species have adopted special storage mechanisms and can store the excess nitrogen as nitrate or in nitrogen-rich amino acids as arginine. At still higher internal nitrogen concentrations, $c_{N,tox}$, toxicity appears. The transition from sufficiency to toxicity is not well known and of little interest in this context. Experimental evidence also indicates that toxicity is reached only when external concentrations attain very high levels,[6] otherwise the plant's internal mechanisms normally stop the uptake at some internal nitrogen concentration between $c_{N,opt}$ and $c_{N,tox}$; uptake through leaves could be a mechanism that does not operate properly in this respect. In this treatise, our interest is in the response range with some digressions into sufficiency. Above and below these ranges the plants are dead and could possibly serve as inputs of litter to the decomposition process.

A consequence of (8.3) is that if either the relative growth rate or the internal nitrogen concentration is constant in time, then the other also has to be so. This implication is also valid under more general models than (8.2), e.g. if P_N is a function of N/W. A corollary to this implication is that if the internal nitrogen concentration, c_N, is constant, then

$$\frac{dc_N}{dt} = \frac{d}{dt}\frac{N}{W} = \frac{1}{W}\frac{dN}{dt} - \frac{N}{W^2}\frac{dW}{dt} = \frac{N}{W}\left(\frac{1}{N}\frac{dN}{dt} - \frac{1}{W}\frac{dW}{dt}\right) = c_N(R_N - R_W) = 0$$

(8.4)

Thus, to maintain a constant plant nutrient concentration, the nitrogen amount in the plant has to increase with a relative rate of R_N equal to the relative growth rate ($R_W = R_N$). From (8.3) follows that R_W is constant and the change in nitrogen amount is then also exponential. We can therefore conclude that only under exponential growth can the nitrogen concentration in the plant be constant, and vice versa, only plants growing at exponential rates have constant nitrogen concentrations.

It is quite natural that a relation such as (8.4) should hold, because a plant growing at an exponential rate and not supplied with nitrogen at the same exponential rate would very rapidly move either into the toxicity or deficiency range as a purely mathematical consequence of the speed with which exponential functions cause changes.

Problem 8.1
Analyse the consequences on plant nitrogen concentration for plants growing with $R_W = 0.05, 0.10, 0.25$ d^{-1} and supplied with nitrogen at a linear rate and for plants with $R_W = R_N$.

8.2   Nitrogen productivity or photosynthesis and respiration

Is it reasonable to expect that such a complicated process as plant growth can be reduced to a single parameter? Do we not need to use models that account for many of the subprocesses involved in growth? In this section, we shall show why plant properties aggregate into the nitrogen productivity by calculating plant carbon balances from photosynthesis minus respiration.[7]

Consider a plant consisting of a leaf biomass, W_L, a root biomass, W_R, and with leaf area S. The size of the plant is $W = W_L + W_R$ and the leaf weight ratio is f_L ($= W_L/W$). Let the net photosynthetic rate per unit leaf area be A. The respiration rate of non-photosynthesising tissue (and this includes leaves in the dark) consists of a maintenance component R_m (d^{-1}) and a growth component R_g (dimensionless). When the dark period of the day is f_d (dimensionless), the relative growth rate of the plant is

$$R_W = \frac{1}{W}\frac{dW}{dt} = A\frac{S}{W_L}f_L(1-f_d)-(R_m + R_g R_W)(1-f_L)-(R_m + R_g R_W)f_L f_d$$

(8.5)

The rate of photosynthesis can be calculated from the Farquhar-von Caemmerer model,[8] i.e.

$$A = \min(V_{cmax}\frac{c_i - k_1}{c_i + k_2}, J_{cmax}\frac{Q}{Q+2.1J_{cmax}}\frac{c_i - k_1}{4.5c_i +10.5k_1}),$$ (8.6)

where V_{cmax} is the maximum rate under Rubisco limited conditions, J_{cmax} the maximum rate under electron transport limited conditions, Q the photon flux density, c_i the CO_2 concentration in the leaves, and k_1 and k_2 parameters.[9]

The respiration rate is calculated with the equation suggested by Williams et al. (1987) to estimate growth respiration

$$R_g = k_{0g} + k_{1g}c_N$$ (8.7)

and the equation by Ryan (1991) to estimate maintenance respiration

$$R_m = k_{0m} + k_{1m}c_N$$ (8.8)

where the ks are parameters.

The variations in leaf weight ratio and specific leaf area with nitrogen availability are expressed as linear functions of c_N

$$f_L = k_{0L} + k_{1L}c_N \tag{8.9}$$

$$\frac{S}{W_L} = k_{0A} + k_{1A}c_N . \tag{8.10}$$

There are only small effects of light and CO_2 on f_L at constant c_N, but the parameters k_{0A} and k_{1A} have to be estimated separately for each new combination of light and CO_2.[10]

We also express J_{cmax} and V_{cmax} as linear functions of c_N[11]

$$J_{cmax} = k_{0J} + k_{1J}c_N \tag{8.11}$$

$$V_{cmax} = k_{0V} + k_{1V}c_N . \tag{8.12}$$

Since most studies of photosynthetic rates are done on "fully developed leaves" there is also a need to account for the development of the photosynthetic performance during the maturation of the leaves. There is no generally accepted model to describe this maturation but empirical evidence[12] suggests that

$$1 - e^{-k_{lm}\tau} \tag{8.13}$$

may be a suitable form to describe how the photosynthetic performance increases with leaf age τ (d) and where k_{lm} (d^{-1}) is the rate of maturation. The physically observed leaf area of the whole plant, $S(t)$, should then be replaced by a photosynthetically effective leaf area, which is weighted for the differences in performance. For an exponentially growing plant we get

$$S_E(t) = \int_0^t R_W S_0 e^{R_W \tau}(1 - e^{-k_{lm}(t-\tau)})d\tau =$$

$$= S_0 e^{R_W t}\left[1 - e^{-R_W t} - \frac{R_W}{R_W + k_{lm}}\left(1 - e^{-(R_W + k_{lm})t}\right)\right] \tag{8.14}$$

$$\xrightarrow[t \to \infty]{} S_0 e^{R_W t}\frac{k_{lm}}{R_W + k_{lm}} = S(t)\frac{k_{lm}}{R_W + k_{lm}}$$

where S_0 is the initial leaf area. Since we will only be concerned with studies where $R_W t \gg 1$, it is sufficient to use only the expression in the last line. Therefore, when k_{lm} is of the same magnitude as R_W, leaf maturation is an important factor to take into account when estimating whole-plant photosynthesis. There is, on the other hand, no need to worry about the senescence of the leaves, because in exponentially growing plants, old leaves will always constitute only a small fraction of the total leaf biomass. The equation for R_W is then

$$R_W = A \frac{S}{W_L} \frac{k_{lm}}{R_W + k_{lm}} f_L (1 - f_d) - (R_m + R_g R_W)(1 - f_L) - (R_m + R_g R_W) f_L f_d$$

(8.15)

It is clear that (8.15) will not give a linear relationship between R_W and c_N; the question is, how good an approximation is the linear relationship? Using Ågren's (1996) parameter estimates for a series of experiments on birch, we can compare (8.15) with (8.3). At least under the conditions of these experiments it is clear that the linear approximation of the nitrogen productivity is satisfactory, Figure 8.2. Indeed, with the errors in the 16 parameters required when calculating plant growth from photosynthesis and respiration, it is likely that, for many years to come the nitrogen productivity will not only be the simplest, but also the most accurate way of doing this.

8.3.   Different uptake models

In this section we will use different explicit uptake models to derive growth curves and dose-response curves corresponding to different uptake functions. We will consider both the entirely general case as well as two cases that should bracket any conceivable uptake function - exponential uptake and no uptake.

8.3.1   *General uptake rate*

Let us first consider the general case where the amount of nitrogen in the plant varies with time according to some general function g

$$N(t) = N_0 g(r_m t, U)$$ (8.16)

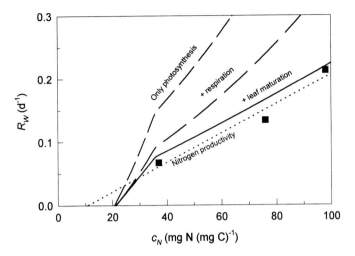

Figure 8.2 Relation between relative growth rate and whole-plant nitrogen concentration calculated from photosynthesis (broken line) and with different corrections (broken line and solid line), from the nitrogen productivity (dotted line), and from (■) measurements by Pettersson & McDonald (1994) (from Ågren 1996).

where N_0 is the initial amount of nitrogen in the plant and r_m the maximum relative growth rate. The variation in uptake with time has two characteristics, which affect plant growth in fundamentally different manners: (i) the functional form of the variation, which is given by g; and (ii) the level of the uptake, which is set by the parameter U, Figure 8.3. For example, g can be a constant, a linear function or an exponential function, in which cases U represents the constant value, the rate of increase, or the relative rate of increase, respectively. We have preferred to express g as a function $r_m t$ rather than just t, because in that way time will never appear explicitly as a dimension in g. Since g must be dimensionless, it is also possible to let U be a dimensionless quantity.

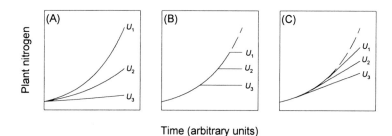

Figure 8.3 Different uptake functions to illustrate the difference between functional form and level of uptake. (A) Different exponential rates. (B) Different fixed amounts. (C) Different linear uptake rates. The broken line in (B) and (C) corresponds to the maximum exponential uptake rate.

The formal solution to (8.1) with P_N constant and $N(t)$ given by (8.16) is then

$$W(t) = W_0 e^{-a't} + P_N N_0 \int_0^t g(r_m \tau, U) e^{a'\tau} d\tau =$$

$$= e^{-r_m t \frac{a'}{r_m}} \left[W_0 + \frac{P_N N_0}{r_m} \int_0^{r_m t} g(z, U) e^{za'/r_m} dz \right] \qquad (8.17a)$$

$$\rightarrow W_0 \left[1 + \frac{R_{W,0}}{r_m} \int_0^{r_m t} g(z, U) dz \right], \text{ when } a' = 0 \qquad (8.17b)$$

where

$$a' = P_N c_{N,min} \qquad (8.18)$$

and

$$R_{W,0} = P_N c_{N,0} = P_N \frac{N_0}{W_0} \qquad (8.19)$$

Hence, if we neglect $c_{N,min}$ (or equivalently a') we see from (8.17) that all growth curves can be described by families of curves characterised by the uptake function g and parameterised by the level of uptake U (dimensionless) and an initial condition expressed in terms of the initial relative growth rate ($R_{W,0}$) divided by the maximum relative growth rate (r_m). Thus, (8.17) shows that plants growing under the same uptake conditions, the same g, will follow the same growth curve independently of species or other growth conditions when time is scaled to r_m.

We can generalise the expression for the relation between supplied amount of nitrogen and growth still more by normalising plant sizes and plant nitrogen amounts. The idea of normalisation is not new and is commonly and with great success used in, e.g. thermodynamics, where the variables thus formed are called reduced variables. The point is that if a variable is divided by (reduced with respect to) some carefully chosen value (in thermodynamics generally a critical point) expressions will simplify by removing parameters referring to the specifics and by retaining the structure of the general. The state of an optimally growing plant can, in our situation, serve as the proper reducing variable. As can be seen in Figure 8.1, such a plant grows at a distinct break-point between two response ranges. Hence, (8.16) and (8.17) are replaced by

$$N_r(t) = \frac{N(t)}{N_{opt}(t)} = \frac{N_0}{N_{0,opt}} e^{-r_m t} g(r_m t, U) \qquad (8.20)$$

$$W_r(t) = \frac{W(t)}{W_{opt}(t)} = \frac{W_0}{W_{0,opt}} e^{-r_m t (1 + a'/r_m)} \left[1 + \frac{R_{W,0}}{r_m} \int_0^{r_m t} g(z,U) e^{za'/r_m} dz \right] \rightarrow$$

$$\rightarrow \frac{W_0}{W_{0,opt}} e^{-r_m t} \left[1 + \frac{R_{W,0}}{r_m} \int_0^{r_m t} g(z,U) dz \right], \quad \text{when } a' \rightarrow 0 \qquad (8.21)$$

where the subscripts r stand for reduced.

Additional simplifications of (8.20) and (8.21) can be achieved by assuming all plants to start from optimal conditions, i.e. $W_0 = W_{0,opt}$, $N_0 = N_{0,opt}$ and consequently $R_{W,0} = r_m$. It is then possible to formally eliminate U between (8.20) and (8.21) to get an expression of the form

$$W_r(t) = f\{N_r(t), r_m t; g\} \qquad (8.22)$$

Hence, the dose-response curves that obtain have only one species-specific and environmentally related parameter, r_m. Otherwise (8.22) holds whatever species and whatever environmental conditions are used. In addition, for certain and important classes of gs, the elimination of U will at the same time eliminate $r_m t$. In these instances, (8.22) becomes a very general result.

Namely, there exists one universal dose-response curve applying to all species and all environmental conditions, subject only to the constraint that P_N can be regarded as constant and $c_{N,min}$ can be neglected. As can be seen in Table 9.1, the latter constraint is often satisfied, except for the most slowly growing plants; or more precisely, a' should be small compared to R_W.

8.3.2   *Exponential uptake*

An exponential uptake can be obtained either from a deliberately exponential supply, in which case the rate can be controlled, or when the uptake is proportional to plant size (or the growth rate of the plant), i.e.

$$\frac{dN}{dt} = kW \qquad (8.23)$$

Using (8.2) and (8.23) and differentiating (8.1) once more gives

$$\frac{d^2W}{dt^2} + a'\frac{dW}{dt} - P_N kW = 0 \qquad (8.24)$$

from which it follows that the plant will grow exponentially at a rate (the corresponding negative root accounts for the adjustment to the exponential growth)

$$\sqrt{P_N k + \frac{a'^2}{4}} - \frac{a'}{2} \rightarrow \sqrt{P_N k} \text{ for } a' \rightarrow 0 \qquad (8.25)$$

Problem 8.2.
Explain why $R_W \sim P_N^{1/2}$ according to (8.25) and $R_W \sim P_N$ according to (8.3) do not contradict each other.

When uptake is exponential we can choose $U = R_N/r_m$ and g is thus given by

$$g(r_m t, U) = e^{Ur_m t} \qquad (8.26)$$

Inserting (8.26) in (8.17) and evaluating the integral gives

$$W(t) = \frac{P_N}{Ur_m + a'} N_0 e^{Ur_m t} + \left(W_0 - \frac{P_N}{Ur_m + a'} N_0\right) e^{-a't} \qquad (8.27)$$

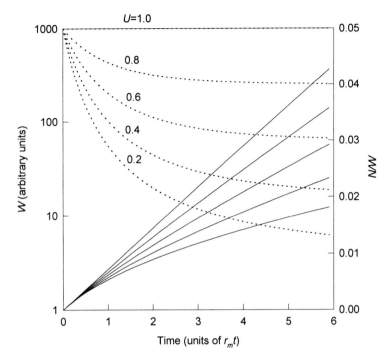

Figure 8.4 Development in time of plant weight (solid lines) and nitrogen concentration (broken lines) for plants supplied with different exponential rates of nitrogen but all starting at optimal concentrations. The two sets of curves are ordered in the same way.

If, for simplicity, we assume that the initial nitrogen concentration in the plants is in "balance" such that the initial transients disappear ($N_0/W_0 = (Ur_m + a')/P_N$), then the plant weight follows exactly the same function as the plant nitrogen, because the parenthesis in front of $e^{-a't}$ disappears.[13] Figure 8.4 shows some curves of plant size and plant nitrogen concentration development for plants starting from optimum conditions.

With the initial conditions defined in such a way that $N_0 = N_{0,opt}$, the nitrogen and weight curves expressed in reduced variables (8.20 and 8.21) become

$$N_r(t) = e^{r_m t(U-1)} \qquad (8.28)$$

$$W_r(t) = \frac{P_N}{Ur_m + a'} \frac{N_0}{W_0} e^{r_m t(U-1)} + \left[1 - \frac{r_m + a'}{Ur_m + a'}\right] e^{-r_m t(1 + a'/r_m)} \qquad (8.29)$$

U can now be solved from (8.28) and inserted in (8.29) to give the dose-response curve

$$W_r = \frac{zN_r + e^{-z} \ln N_r}{z + \ln N_r} \rightarrow N_r \text{ when } z \rightarrow \infty \qquad (8.30)$$

where

$$z = (1 + a'/r_m)r_m t \qquad (8.31)$$

Alternatively, the dose-response curve is not expressed in terms of the reduced amount of nitrogen in the plant, but in terms of the reduced nitrogen concentration in the plant

$$c_{N,r} = \frac{c_N}{c_{N,opt}} = \frac{N/W}{(N/W)_{opt}} = \frac{N_r}{W_r} \qquad (8.32)$$

giving the implicit equation

$$W_r = \frac{zW_r c_{N,r} + e^{-z} \ln W_r + e^{-z} \ln c_{N,r}}{z + \ln W_r + \ln c_{N,r}} \qquad (8.33)$$

These dose-response curves will change with the length of the growth period (z depends on t). Initially ($z = 0$), the curve is the horizontal line $W_r = 1$, reflecting the choice of initial conditions; growth is started from optimal conditions. With time the curves will swing downwards to asymptotically approach the straight line $W_r = N_r$. Figure 8.5 illustrates this process.

Problem 8.3.
Derive (8.30) and (8.33).

8.3.3   *Fixed amount*

By this label we understand a situation where the plant is given a predefined amount and once this amount is exhausted no external supply of nitrogen exists and the plant is directed to live on its internal resources. We

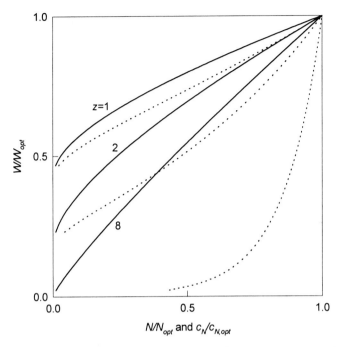

Figure 8.5 Dose-response curve in reduced variables under exponential nitrogen uptake and for various growth periods. Solid lines $= N_r$. Broken lines $= c_{N,r}$.

idealise the uptake curve in this case, in such a way that the uptake proceeds at the optimal rate until all nitrogen in the external medium is taken up (plant nitrogen amount is N_{max} at this point), whereafter the uptake is zero, cf. Figure 8.3B. Such idealisation is reasonable in view of what short-term uptake experiments tell us. As has been repeatedly shown since Olsen's (1950) early experiments, the uptake rate of a plant can remain essentially constant until very small amounts remain in the root medium, given that large diffusive resistances are avoided.[14]

The function g in this case is defined by

$$g(r_m t; U) = \min\{e^{r_m t}, U\} \qquad (8.34)$$

where $U = N_{max}/N_0$. The plant weight increases exponentially at the maximum rate r_m up to some time t_b when all external nitrogen has been taken up, and

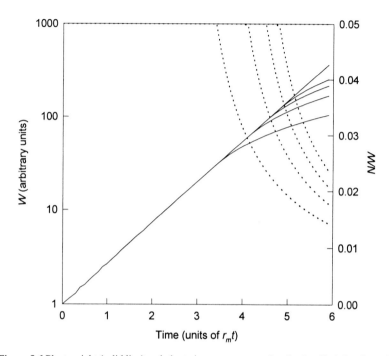

Figure 8.6 Plant weight (solid line) and plant nitrogen concentration (broken line) for plants that have been given different fixed amounts of nitrogen.

from then on the growth is approximately linear until the internal concentration approaches $c_{N,min}$ (or as long as $a'(t - t_b) \ll 1$), when growth follows a negative exponential function

$$W(t) = W_0 e^{r_m t}, \quad t \le t_b \tag{8.35}$$

$$W(t) = W(t_b)\left[1 + (1 - e^{-a'(t-t_b)})\frac{r_m}{a'}\right], \quad t > t_b$$
$$\rightarrow W(t_b)\left[1 + r_m(t - t_b)\right] \text{ for } a' \rightarrow 0 \tag{8.36}$$

where

$$e^{r_m t_b} = U \tag{8.37}$$

Figure 8.6 shows some examples of such growth curves.

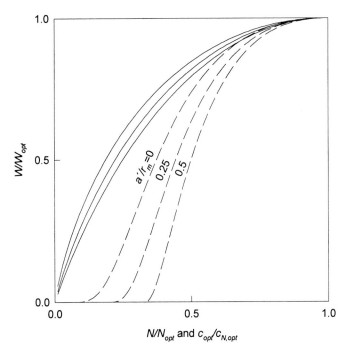

Figure 8.7 Dose-response curves in reduced variables for plants given a fixed amount of nitrogen for different values of a'/r_m. Solid line = N_r. Broken line = $c_{N,r}$.

For plants that have grown for a shorter time than t_b, the growth response curves are those derived in the previous section for exponential additions. For plants that all have grown for longer than t_b, the following response curves can be derived

$$W_r = N_r \left[1 + (1 - N_r^{a'/r_m}) \frac{r_m}{a'} \right]$$
$$\rightarrow N_r (1 - \ln N_r) \quad \text{for } a' \rightarrow 0 \tag{8.38}$$

$$N_r = Ue^{-r_m t} = e^{-r_m(t - t_b)} \tag{8.39}$$

Alternatively, we express W_r in terms of the reduced nitrogen concentration

$$W_r = \frac{\left[1 - \left(\dfrac{1}{c_{N,r}} - 1\right)\dfrac{a'}{r_m}\right]^{r_m/a'}}{c_{N,r}} \rightarrow \frac{e^{1/c_{N,r}-1}}{c_{N,r}} \quad \text{for } a' \rightarrow 0 \qquad (8.40)$$

Figure 8.7 shows these dose-response curves. In the physiological literature such curves are considered as the normal dose-response curves,[15] although the analysis here shows that the shape of the dose-response curve is strongly linked to the way the nutrient is supplied to the plant, i.e. to function g. It should also be observed that when a' can be neglected, (8.38) contains no parameters describing either plant properties or environmental conditions. This dose-response curve holds universally.

Problem 8.4.

Derive (8.38) and (8.40).

Problem 8.5.

Compare the solutions (8.35) and (8.36) with the growth curves obtained from the equations $dW/dt = K_1 W/(K_2 + W)$ (Greenwood et al. 1991).

Problem 8.6.

In plant nurseries it is desirable to obtain plants with sizes that are as equal as possible. (A) How will a distribution of plant sizes change if you distribute nutrients homogeneously over the area where the plants are growing? (B) How should you distribute the nutrients if you want the distribution to narrow with time?

8.3.4 *General uptake rate - revisited*

The two special cases considered above can be seen as the extremes of a continuum of uptake functions. The exponential uptake at maximum rate is the fastest possible uptake rate whereas the fixed amount turns into a situation where the uptake rate is zero. A reasonable series of uptake rates can be generated by requiring that higher and higher derivatives are continuous in the uptake function. The fixed amount is then the case when only the zeroth order derivative (the function itself) is continuous and the maximum exponential uptake the case where all derivatives are continuous. The function g can then be a given as a truncation of the Maclaurin expansion of the exponential function (U is the number of terms in the series)

$$g(r_m t; U) = 1 + r_m t + \frac{1}{2}(r_m t)^2 + \frac{1}{2 \cdot 3}(r_m t)^3 + ... + \frac{1}{U!}(r_m t)^U \qquad (8.41)$$

In the simplest case, where $a' = 0$ and the plants start at optimal conditions $(P_N N_0 = r_m W_0)$, (8.17b) gives

$$W(t) = W_0\left[1 + r_m t + +\frac{1}{2}(r_m t)^2 + ... + \frac{1}{(U+1)!}(r_m t)^{U+1}\right] \qquad (8.42)$$

i.e. the plant weight forms a similar series. It should be observed that plant weight has one order higher continuous derivatives than nitrogen amount. The significance of this is that it shows how transitions between truly exponential growth and non-exponential ones occur. Even in the most clear-cut case where merely the zeroth order derivative of the uptake rate is continuous ($U = 0$), not only the plant weight but also its first derivative are continuous in time and only the second derivative will show a discontinuity. The possibility of demonstrating such discontinuities requires data of very high quality. Claims that growth has been exponential and which are based on plant weight increases must therefore be treated with great caution.

8.4   Nitrogen productivity and other nutrients

Can the concept of nitrogen productivity be generalised to cover situations where other mineral nutrients also become limiting? The famous law by de Saussure and Sprengel, but more widely known under the name of Liebig's law of the minimum[16] states that it is the factor in minimum that will determine growth rate. If we restrict its use to only mineral nutrients we have a possible generalisation. Other factors such as light and temperature limit growth simultaneously but independently of the mineral nutrient elements. We like to think of them as orthogonal factors and will discuss them further below. We introduce a general nutrient productivity concept[17] and assume that (8.1) and (8.2) apply to any mineral nutrient. The growth rate is set by the combination of productivity and amount of nutrient giving the lowest growth rate, i.e.

$$\frac{dW}{dt} = \min_i \left\{P_i(n_i - c_{i,min}W)\right\} \qquad (8.43)$$

where the minimum is taken over all nutrients i. To see the consequences of (8.43), let us apply it to nitrogen and one other nutrient (i) with productivities P_N and P_i, respectively. (8.3) is then replaced by

$$R_W = \min\left\{ P_N(c_N - c_{N,min}), P_i(c_i - c_{i,min}) \right\}$$

$$= \min\left\{ P_N(c_N - c_{N,min}), P_i(\frac{c_i}{c_N}c_N - c_{i,min}) \right\} \qquad (8.44)$$

Thus, as long as nutrient i is supplied in sufficient amounts, there is no difference relative to our previous discussions. With nutrient i in the response range, the relative growth rate is a linear function of c_i, and growth can be expressed with the productivity of nutrient i as base. However, if nitrogen and nutrient i occur in constant proportions, the relative growth rate can again be expressed as a linear function of the nitrogen concentration but with a slope (the nitrogen productivity) proportional to the ratio between the concentration of nutrient i and nitrogen. This makes it possible to keep nitrogen as the basis and let other nutrients just modify the nitrogen productivity.

The nutrients to which (8.44) should apply are those which are quantitatively and directly involved in the growth process. Nutrients, which themselves are not directly involved in growth, but stimulate the functioning of other nutrients, require different formulations. Consider a nutrient k, the effect of which is to regulate the productivity of nutrient i (orthogonal nutrients). This leads to different consequences from (8.44), which we can illustrate by assuming that P_i is proportional to the concentration of nutrient k.

$$P_i = P_{ik}c_{ik} \qquad (8.45)$$

The relation between the relative growth rate and the concentration of nutrient i is then

$$R_W = P_{ik}c_k(c_i - c_{i,min}) = P_{ik}\frac{c_k}{c_i}(c_i - c_{i,min})c_i \qquad (8.46)$$

Hence, when the ratio between c_k and c_i is constant, the relative growth rate is no longer a linear function of the concentration of nutrient i but a quadratic one. It should be possible experimentally to distinguish between the consequences in (8.44) and (8.46) of the two different hypotheses (8.43) and (8.45). Experiments with phosphorus and *Betula pendula*[18] and *Pinus sylvestris*[19] show that phosphorus should be treated in a similar way as nitrogen, e.g. following (8.44).

Minerals like K, Ca, Mg, Fe and some others are more likely to belong to the second category. When supplied in limiting amounts they behave according to (8.44), see also Figure 9.2, but we are not aware of any test that would distinguish between (8.44) and (8.46). Differences between the way elements affect allocation are, however, evident.[20],[21] The discussion throughout has been around the effects of nutrients on growth. However, several mineral nutrients have additional important effects (e.g. potassium as a protector against pathogens). Figure 8.8 shows an example of how such effects can also be analysed with the aid of the nutrient productivity concept.

8.5 Variable P_N

Let us use (8.3) and (8.4) to evaluate some qualitative aspects of plant growth when P_N is changed as a result of changes in other environmental variables. Such changes can easily lead to the counter-intuitive consequence that plant size becomes negatively correlated with plant nutrient concentration. For simplicity we set $c_{N,min} = 0$. We can represent (8.3) and (8.4) in a single diagram with R_W as common axis, Figure 8.9. Start in the figure at the optimal nitrogen concentration on the c_N-axis and follow the dotted lines up to the intersection with one of the solid lines representing (8.3), A or D. Turn left and continue, crossing the R_W-axis, until the intersection with the solid line representing (8.4), B or B'. Move down from there to the R_N-axis. We see that depending upon the value of P_N we end up at different values of R_N; the higher the value of P_N, the higher will be R_N. For a plant to be able to utilise a high nitrogen flux density (high R_N) it is necessary that it has a high nitrogen productivity.

A plant growing at a given R_N has a fixed relative growth rate and an increase in nitrogen productivity cannot increase it. The absolute growth rate is, however, increased, (8.1), and as a consequence the nitrogen concentration decreases; compare points C and A or C' and A'. Under these circumstances decreased concentration of a nutrient is a sign of improved growth conditions. Decreased nitrogen concentration as a result of moving down a line of constant P_N is, on the other hand, a sign of decreasing nitrogen availability and therefore of less favourable growth conditions; compare C and C' or A and A'.

Next, we will consider groups of plants with initially the same amount of nitrogen, the same nitrogen concentration, and supplied with nitrogen according to the same function g but growing under otherwise different conditions. This would be a typical set-up for studies of, e.g. different light intensities or carbon dioxide concentrations. Let us also, for simplicity, again set $c_{N,min} = 0$. Then with (8.17),

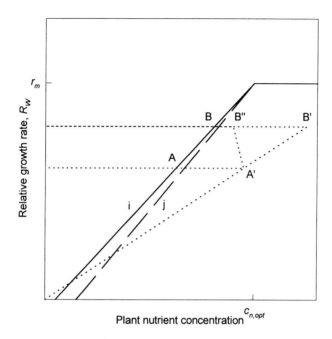

Figure 8.8 Hypothetical relationship between R_W and c_n for two nutrients, i (solid line) and j (broken line), respectively. Additional, dotted, lines are inserted to facilitate the interpretation. Assume that a constant R_W is maintained at level A and that nutrient i is growth-limiting. The concentration of i is then the one corresponding to point A. Nutrient j, being non-limiting, must be at a concentration somewhere along the dotted line to the right of its intersection with the broken line, e.g. A'. Suppose now that R_W is increased up to level B through an increased, but still limiting, supply of nutrient i. If nutrient j is taken up in the same proportions at both levels A and B, the concentration of j will move up to the point indicated by B'. If, on the other hand, the supply of j is not correspondingly adjusted, the concentration of j may end up at level B". If the amount of j not directly involved in growth (the amount of which is given by the difference between the actual concentration and the broken, growth-limiting line) has some other function, the change of supply of nutrient i might have consequences on plant performance that are not related to the level of i per se but are indirect effects caused by a decreased "excess" of other nutrients (from Ingestad & Ågren 1992).

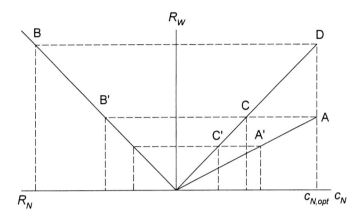

Figure 8.9 Relations between R_N, R_W, and c_N at different P_N. $c_{N,min} = 0$.

$$W(t) = W_0 + P_N N_0 \int_0^t g(\tau)d\tau = W_0 + P_N N_0 G(t) \qquad (8.47)$$

The relative growth rates and nitrogen concentrations of these plants can then be written

$$R_W = \frac{1}{W}\frac{dW}{dt} = \frac{P_N N_0 g(t)}{W_0 + P_N N_0 G(t)} \qquad (8.48)$$

$$c_N = \frac{N}{W} = \frac{N_0 g(t)}{W_0 + P_N N_0 G(t)} \qquad (8.49)$$

If P_N is eliminated between (8.48) and (8.49) we get

$$R_W = \frac{g(t)}{G(t)} - \frac{W_0}{N_0 G(t)} c_N \qquad (8.50)$$

Since $g(t)$ (and hence $G(t)$) is the same for all plants (8.50) describes a straight line that connects these plants in the R_W-c_N plane, Figure 8.10; (8.50) is the general formulation of what Linder & Rook (1984) showed as a special case. The slope ($W_0/N_0 G(t)$) and intercept ($g(t)/G(t)$) change, however, with time. The intercept with the c_N-axis ($N_0 g(t)/W_0$) increases with time, which is why plants with a sufficiently low P_N must initially increase their c_N and hence relative growth rate. For plants with high P_N the effect of the decreasing magnitude of the slope dominates and they have monotonically decreasing relative growth rates. The negative slope in (8.50) ensures, however, that the plants starting at the highest relative growth rates will maintain the highest relative growth rates although the differences in relative growth rates are decreasing. Simultaneously, the ranking with respect to internal nitrogen concentrations is retained, but with the slowest growing plants at the highest concentrations. The long-term consequence of a limited nitrogen supply is expressed as a change from ranking the different plants according to relative growth rates at equal internal nitrogen concentrations to ranking according to internal nitrogen concentrations at equal relative growth rate.

Problem 8.7.
Show in the special case of exponential uptake that (8.50) reduces to $R_W = R_N$.

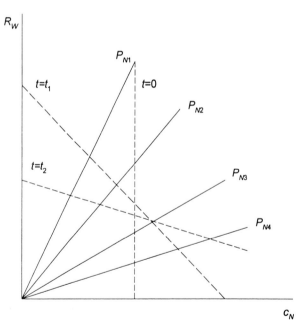

Figure 8.10 Relation between relative growth rate and plant nitrogen concentration for plants initially at the same concentration and being supplied according to the same uptake function but having different nitrogen productivities. $P_{N1} < P_{N2} < P_{N3} < P_{N4}$.

8.6   Comparison with other approaches

There are alternative ways of calculating the carbon fixation by a plant or stand. The two most commonly used are based on estimates of (net) photosynthesis (e.g. Ågren et al. 1991, see also section 8.2) or light use efficiency (Monteith 1977, Linder 1985). The use of photosynthesis links carbon fixation directly to the underlying process with ample opportunities to relate to more detailed levels of understanding. The price paid is that on a temporal scale it is necessary to work at a time resolution of less than a day; time averages of photosynthesis over longer periods lack simple biological interpretations. In contrast, with nutrient productivities it is the rate of variability in the plant's nutrient amount that limits the temporal scale. Under laboratory conditions, nutrient amounts typically change significantly over days whereas under field conditions changes can take place over years. Hence the ease with which the same nutrient productivity concept can be moved between so different conditions.

Light use efficiency and nutrient productivities are similar in the sense that both integrate a large number of complicated processes and reduce them to a simple concept. Their use also requires about the same amount of information (one plant and one external variable); plant leaf area (LAI) and incoming light or plant nutrient content and climate. McMurtrie et al. (1994) used light use efficiency to provide an elegant analysis of climatic effects on plant growth. However, plant leaf area cannot reliably be calculated without reference to plant nutrient status. It should therefore be easier to choose a way that directly addresses this point, i.e. through nutrient productivities.

NOTES

[1] Baule (1956).
[2] e.g. Duncan et al. (1967).
[3] Ågren et al. (1980), Ågren & Bosatta (1990). The early book *Mathematical Models in Plant Physiology* by Thornley (1976) and later books by Charles-Edwards (1981) and Landsberg (1986) with largely analytical approaches have not been enough to persuade the scientific community of modellers of their value.
[4] e.g. Legendre & Legendre (1983).
[5] Following Gauch (1972) we have indicated a series of response ranges.
[6] Ingestad & Lund (1979; Ingestad (1982).
[7] This section is a summary of Ågren (1996).

[8] Farquhar & von Caemmerer (1982).

[9] This is not quite accurate as the Farquhar-von Caemmerer model requires that leaf respiration during day-time should be calculated separately from photosynthesis. However, the dark respiration can be difficult to estimate separately because of its change with the length of time in darkness (Byrd et al. 1992). We assume that its effects can be absorbed in the parameter estimates. Equation (8.6) is probably accurate enough for the present purpose.

[10] Pettersson et al. (1993).

[11] Field (1983), Pettersson & McDonald (1994).

[12] Šesták (1985).

[13] Note, however, that this choice of initial conditions means that plants with a smaller U will initially be heavier if they are to contain equal amounts of nitrogen. The alternative initial conditions, equal plant weights, lead to different initial nitrogen amounts in the plants, and equal initial nitrogen concentrations will involve an uninteresting transient as the nitrogen concentrations asymptotically approach steady state.

[14] cf. Ingestad & Ågren (1988, 1992).

[15] e.g. Epstein (1972).

[16] Liebig (1840).

[17] Ågren (1988).

[18] Ericsson & Ingestad (1988).

[19] Ingestad & Kähr (1985).

[20] Ericsson (1995).

[21] Nutrient productivity is a very general concept, not only applicable to mineral nutrients but also to other essential nutrients such as vitamins, e.g. vitamin B_{12} (Ågren 1988).

9

Plant growth - applications and extensions

In this chapter we will test some of the predictions of the previous chapter against empirical observations. A natural step in this direction is to look at responses to available light and the consequences of light extinction in a canopy. We will also look at the problem of allocation and how it can be handled within the frame-work of nitrogen productivity. A comparison with other approaches to calculations of plant growth concludes the chapter.

9.1   Empirical evidence for nutrient productivities

9.1.1   *Exponential growth*

In principle, any nutrition experiment where plant weights and nitrogen contents have been monitored regularly could serve as a test of the predictions of (8.3) but, as is evident from (8.3) and (8.4), the simplest system to analyse is the exponentially growing plant, although precaution has to be taken to ensure that the observed growth is indeed exponential, cf. (8.42). The safest way to obtain exponential growth is to supply the plants with an exponentially increasing supply of a limiting resource as (8.4) suggests. This methodology has been tested on numerous occasions[1] and Figure 9.1 provides one example.

The results presented in Figure 9.1 are independent of any model for nitrogen productivity and demonstrate only the validity of the experimental technique. A test of the specific model (8.2), P_N constant, is given in Figure 9.2. The agreement between the model and the empirical observations is as good as one can reasonably expect. There are some important differences between species that are evident in Figure 9.2. There are large differences in the maximum relative growth rates and $c_{N,min}$ between the two species. The birch also seems to use all the nitrogen taken up, with no signs of excess

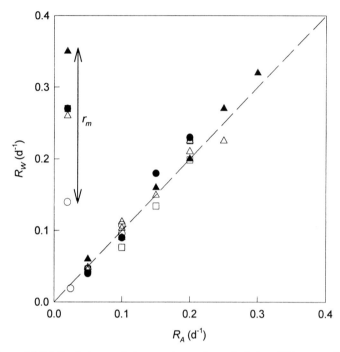

Figure 9.1 Relation between relative growth rate, R_W, and relative addition rate of nitrogen, R_A, in experiments with tomatoes (solid symbols) and birch (open symbols). The experiments have been conducted at different light intensities ($\bullet < \blacktriangle$ and $\circ < \triangle < \square$) and the maximum achievable relative growth rate increases with increasing light intensity). Data from Ingestad et al. (1994ab).

accumulation. In response to increasing light, the birch increases its nitrogen productivity but the major response of the tomato seems to be a decreased $c_{N,min}$. This observation requires, however, further empirical testing before it can be definitely concluded. A more extensive test of nutrient productivity is given in Figure 9.3 where the relationship between R_W and c_n is summarised for several different species and with N, P, K, S, Mg, Mn, Fe, Zn, and vitamin B_{12} as limiting elements.

Table 9.1 summarises nitrogen productivities and parameter values for some species for which this type of experiment has been performed. The statistical significance is generally high. In cases where the parameters are calculated from few points these cover the entire response range and additional experimental points would not, considering the experience of this type of experiment, change the regression lines. The absolute values of the parameters will depend upon the environmental conditions and the values in Table 9.1 would therefore be different if obtained under, for example, other light conditions. However, the values in Table 9.1 are for rather similar environmental conditions and serve to illustrate the variability between species and the ranges of parameters to be expected.[2]

9.1.2   Non-exponential growth

It should be emphasized that the basic equations (8.1) and (8.2) describe instantaneous changes and derivations based on them hold also when the internal nitrogen concentration is changing. We should therefore also test them under non-exponential conditions. There are not many experiments of this category with well-documented supply rates.[3] However, some recent experiments with cabbage and lettuce,[4] where nitrogen additions were interrupted at some point, fit exactly to the model of fixed amounts. Wikström &

Table 9.1 Estimated parameters for a number of species.

Species	P_N gdw(gN)$^{-1}$d^{-1}	r_m d^{-1}	$c_{N,opt}$ mg g^{-1}	$c_{N,min}$ mg g^{-1}	Source
Alnus incana	7.45	0.19	34	8	1
Betula pendula	6.30	0.25	38	4	2
Lemna gibba	17.62	0.56	34	2	3
Lemna minor	12.31	0.65	47	–3	3
Lemna paucicostata	15.78	0.53	36	2	3
Paulownia tomentosa	6.08	0.26	42	0	4
Populus simonii	6.20	0.19	42	12	4
Picea abies, s.p.	4.10	0.071	20	5	5
Picea abies, n.p.	4.86	0.061	19	6	5
Pinus contorta	3.54	0.074	19	1	5
Pinus sylvestris, s.p.	3.07	0.074	22	1	5
Pinus sylvestris, n.p.	3.59	0.073	22	3	5
Lycopersicon esculentum	11.0	0.35	46	15	6

Sources: 1. Ingestad (1980, 1981). 2. Ingestad (1979, 1981). 3. Ericsson et al. (1982). 4. Jia & Ingestad (1984). 5. Ingestad & Kähr (1985). s.p. = southern provenience. n.p. = northern provenience; 6: Ingestad et al. (1994a).

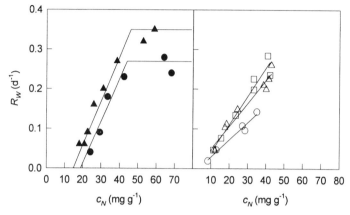

Figure 9.2 Relation between relative growth rate, R_W, and plant nitrogen concentration, c_N, in experiments with tomatoes (solid symbols) and birch (open symbols). The experiments have been conducted at different light intensities ($\bullet < \blacktriangle$ and $\bigcirc < \triangle < \square$) (data from Ingestad et al. 1994ab).

Ågren (1995) showed that these experiments could be explained on the basis of the nitrogen productivities of these plants and with $c_{N,min} = 0$. Expressed in reduced variables, these experiments should therefore fit the universal dose-response curve (8.38), which indeed they do, Figure 9.4.

9.2   Nitrogen productivity and light

One of the factors that we know should affect the magnitude of nitrogen productivity is the light intensity under which plants are growing. Much research has been devoted to describing the relationship between light intensity and photosynthetic rates. A commonly used formula to express this relation is the rectangular hyperbola, cf. (9.1). Work by Ingestad shows that the relation between relative growth rate, plant nitrogen concentration and light intensity can also be given in such a way that the relative growth rate is a linear function of the nitrogen concentration multiplied by a rectangular hyperbola of the light intensity, (9.1) and Figure 9.5.[5] This study suggests that the nitrogen productivity should be described by a rectangular hyperbola of light intensity defined by the maximum nitrogen productivity ($P_{N,max}$), the light intensity (Q), and the light intensity at which the nitrogen productivity is at half its maximum value ($Q_{1/2}$).

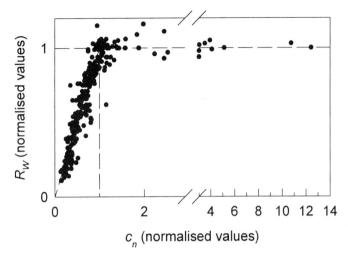

Figure 9.3 Relation between R_W and c_n for 19 different species and with N, P, K, S, Mg, Mn, Fe, Zn, and vitamin B_{12} as limiting elements. Both axes are scaled in such a way that both optimum R_W and c_n are set to 1. The figure contains 282 data points (from Wikström & Ågren 1995).

Thus,

$$P_N = P_{N,max} \frac{Q}{Q + Q_{1/2}} \tag{9.1}$$

Let us now evaluate what will happen to a stand where, because of mutual shading between leaves, the decreasing light intensity with depth in the canopy is important. For simplicity we will consider only horizontally homogeneous canopies. The nitrogen productivity at a given depth in the canopy will then depend upon the local light intensity, $Q(z)$, where z denotes depth counted from the top and downwards in a canopy of total length L. When the nitrogen distribution in the canopy is given by $N(z)$ ($N(z)dz$ is the amount of nitrogen in the interval $[z,z + dz]$), the growth rate of the canopy is

$$\frac{dW}{dt} = \int_0^L P_N(Q(z))N(z)dz \tag{9.2}$$

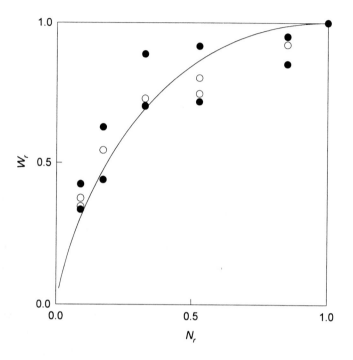

Figure 9.4 Reduced plant mass versus reduced plant nitrogen content for an experiment with cabbage (○) and lettuce (●) and the theoretically predicted relationship (8.38) (line). Data from Burns (1994) and Wikström & Ågren (1995).

Let us further assume that the light intensity decreases according to the Lambert-Beer law with the accumulated amount of nitrogen above the given level in the canopy and with a light extinction coefficient of γ

$$I(z) = I_0 \exp\left\{-\gamma \int_0^z N(\zeta)d\zeta\right\} \tag{9.3}$$

Using (9.1) and (9.2) we can evaluate the integral in (9.3)

$$\frac{dW}{dt} = \frac{P_{N,max}}{\gamma} \ln \frac{Q_0 + Q_{1/2}}{Q_0 e^{-\gamma N} + Q_{1/2}} \tag{9.4}$$

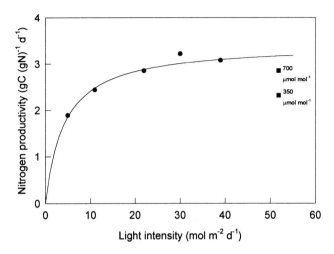

Figure 9.5 Relation between nitrogen productivity and light intensity. Line: Equation (9.1) with $Q_{1/2} = 6.1$. Symbols are experimental data with birch: (●) calculated from Ingestad et al. (1994b); (■) calculated from Pettersson & McDonald (1994), which also includes two CO_2 levels (redrawn from Ågren 1996).

We can simplify (9.4) by expanding it up to third order terms in N

$$\frac{dW}{dt} = P_{N,max} \frac{Q_0}{Q_0 + Q_{1/2}} \left[1 - \frac{\gamma Q_{1/2}}{2(Q_0 + Q_{1/2})} N - \frac{Q_{1/2}\gamma^2(Q_0 - Q_{1/2})}{3(Q_0 + Q_{1/2})^2} N^2 + \cdots \right] N \approx$$
$$\approx (a - bW)N$$

(9.5)

As a first approximation we can describe the effects of decreasing light intensities with depth in a canopy as a linear, decreasing function of the amount of nitrogen in the canopy. However, as light extinction coefficients normally are expressed in units of leaf area or leaf biomass it is more conven-ient to let the decrease be with respect to canopy biomass. This is the reason for W in the last equality in (9.5).

The parameter a in (9.5) has the normal units of nitrogen productivity (e.g. kg dw $(kg N)^{-1}yr^{-1}$). Parameter b, on the other hand, has units such as ha $(kg N)^{-1}yr^{-1}$. With such units it is not likely that parameter b will be of a fundamental nature. The combination a/b, on the other hand, is meaningful with the units kg dw ha^{-1} and can be interpreted as maximum biomass; the growth rate ceases at this biomass. An analysis of the applicability of (9.5) to growth of canopies of conifers was made by Ågren (1983), Figure 9.6. The linear decrease predicted in (9.5) can be seen to hold quite well, when considering that the stands utilised cover large ranges of climatic and other growth conditions. In (9.5) the magnitude of the decrease in nitrogen productivity is derived in terms of light extinction and photosynthetic properties of the species. These parameters have been estimated and can be used in comparison with a/b estimated from the regressions in Figure 9.6. Such a comparison is made in Table 9.2.

It should be emphasized that the linear approximation in (9.5) is done for computational convenience and that other functional forms might be preferable. For example, Smolders et al. (1993) argue from experiments with small spinach plants that the decrease in nitrogen productivity with plant size is better described by $W/(W + parameter)$, and Comins & McMurtrie (1993) use an exponentially decreasing function. The accuracy in data available for

Table 9.2 *The parameters a and b in (9.5) and a comparison between two estimates of the ratio a/b. From Ågren (1983) and Ingestad & Ågren (1984). The first estimate is based on light extinction coefficients and the assumption that $Q_{1/2}$ is the same as the light intensity where photosynthesis is half-maximum, and the second on the values of a and b estimated from the regressions in Figure 9.6.*

Species	a	b	$Q_{1/2}$	$\dfrac{2(Q_0 / Q_{1/2} + 1)}{\gamma}$	$\dfrac{a}{b}$
	kg kg^{-1}yr^{-1}	ha kg^{-1}yr^{-1}	µE m^{-2} s^{-1}	kg ha^{-1}	kg ha^{-1}
Picea abies	18.4	$0.377 \cdot 10^{-3}$	98	34000	49000
Pseudotsuga menziesii	34.4	1.17	258	17000	29400
Pinus nigra	50.2	1.84	178	22000	27000
Pinus resinosa	55.7	2.41	189	21000	23000
Pinus sylvestris	41.4	2.04	330	15000	20000
Salix viminalis	1.34[*]	0.183[*]	356	2800	7300

[*] time unit d^{-1}

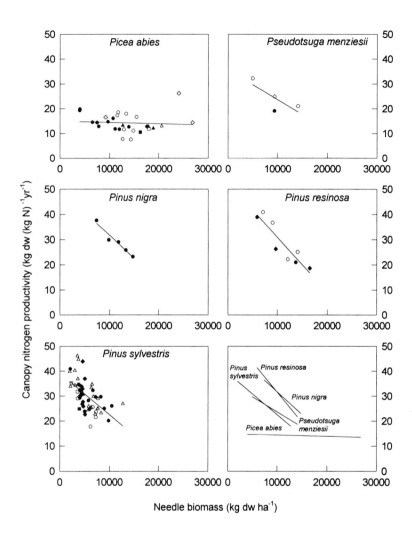

Figure 9.6 Nitrogen productivity for production of new foliage versus needle biomass for some conifers. Different symbols indicate different stands or treatments within one stand (from Ågren 1983). Data (♦) from Mälkönen & Kukkola (1991) added for *Pinus sylvestris* and from Bockheim et al. (1986) for *Pinus resinosa*.

parameter estimation under field conditions is, for the moment, not good enough for us to suggest that one formulation is clearly preferable to the other, although a function that decreases asymptotically to zero might be preferable to one that becomes zero at some finite value.

Problem 9.1.

Add a term $-fW$ to the last equality in (9.5) to account for mortality and calculate how this affects the maximum biomass.

9.3   Root:shoot ratios

The nitrogen productivity concept has so far been applied to whole-plant growth and whole-plant nitrogen amounts, which is reasonable as growth is taking place in the entire plant. There are other growth related processes which, on the other hand, are associated only with a part of the plant. Photosynthesis, for example, is only a shoot process. Hence, by combining growth as related to the whole plant with photosynthesis which is only related to shoots it should be possible to derive an expression for root:shoot ratios. Let, therefore W_L and W_R ($W_L + W_R = W$) be the shoot and root biomass, respectively, and A the rate of assimilation per unit shoot biomass. We can then express plant growth in two ways[6]

$$\frac{d}{dt}(W_L + W_R) = AW_L = P_N N \tag{9.6}$$

We further need an assumption about the variation in photosynthetic rate with nitrogen concentration. The experimental evidence in this area is not reliable as the majority of the experiments have been performed under non-steady nutrition which may obscure their interpretation. Let us first consider a general case

$$A(c_N) = A_m \Phi(c_N) \tag{9.7}$$

where A_m is the maximum net assimilation rate, and $0 < \Phi(c_N) < 1$ is a function to be specified later. From the last equality in (9.6) and (9.7) the root:shoot ratio can then be calculated

$$\frac{W_R}{W_L} = \frac{A_m}{P_N}\frac{\Phi(c_N)}{c_N} - 1 \tag{9.8}$$

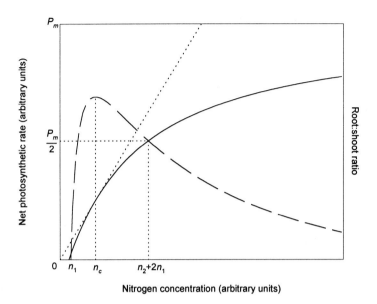

Figure 9.7 Net assimilation rate per unit shoot weight (solid line) and root:shoot ratio (broken line) as functions of plant nitrogen concentration. The curves were drawn using a hyperbolic function for $P(c_N)$.

Denote with Φ' the derivative of Φ with respect to c_N. Taking the derivative (9.8) with respect to c_N we see that with increasing nitrogen concentration, the root:shoot ratio increases if $c_N\Phi'(c_N) - \Phi(c_N) > 0$ and decreases if $c_N\Phi'(c_N) - \Phi(c_N) < 0$. It is now straight-forward to show that the sign of $c_N\Phi'(c_N) - \Phi(c_N)$ is the opposite of the sign of the intercept of the tangent to the assimilation rate-nitrogen concentration curve. Thus, if the intercept is negative, W_R/W_L increases, and if it is positive, W_R/W_L decreases with increasing nitrogen concentration. The qualitatively important aspects of the function $A(c_N)$ are, therefore, given by the sign of the intercept of the tangent to the function.

The general behaviour of (9.7) and (9.8) is shown in Figure 9.7. The characteristic feature of the root:shoot ratio curve is the maximum in the curve occurring at a nitrogen concentration n_c. The maximum in the root:shoot ratio curve appears because at nitrogen concentrations below n_c any further decreases in the nitrogen concentration will decrease the assimilation rate so rapidly that a larger fraction of the plant must be devoted to assimilation in order to maintain the plant's balance between production and consumption of photosynthate. At nitrogen concentrations above n_c, the assimilation rate increases so slowly that when the nitrogen concentration is further increased more shoot biomass is required to supply the plant with the increased demand for photosynthate caused by the increased amount of nitrogen.[7]

Problem 9.2.

Show the statements in the paragraph following (9.8) and derive n_c.

We argued previously that nitrogen productivity should depend upon light according to an hyperbolic function, (9.1). Now, the response of assimilation to light is also generally represented by a hyperbolic function. If the parameters defining the light intensity where these two response curves attain half their maximum values (parameter $Q_{1/2}$ in (9.1) and $2.1J_{cmax}$ in (8.6)) are equal, or nearly so, we then can see from (9.8) that the root:shoot ratio will be independent of, or only weakly dependent, on light.[8]

The root:shoot ratio was derived above as a direct consequence of a functional balance between assimilation and growth determined by the nitrogen content of the plant. One might expect that the root:shoot ratio of a plant should result from an optimisation of resource allocation. Let us see the consequences of assuming that the relative growth rate results from optimising with respect to the root:shoot ratio. From (9.6) we get

$$R_W = A\frac{W_L}{W} - R_d\frac{W_R}{W} = (A + R_d)f_S - R_d \tag{9.9}$$

where we have added a term for root respiration (R_d) and introduced $f_S = W_L/W$. Optimising R_W with respect to f_S means that

$$\frac{dR_W}{df_S} = A + R_d + f_S\frac{dA}{df_S} = 0 \tag{9.10}$$

Since A, R_d, and f_S are positive quantities it follows from (9.10) that

$$\frac{dA}{df_S} < 0 \tag{9.11}$$

It is easier to see the implications of (9.11) if it is rewritten in terms of plant nitrogen concentration. Thus, we write (9.11) as

$$\frac{dA/dc_N}{df_S/dc_N} < 0 \tag{9.12}$$

The general empirical experience is that both dA/dc_N and df_s/dc_N are positive. However, to satisfy (9.12) these two terms need to be of different signs. Our conclusion is therefore that it is not possible to find a solution to (9.11) that is consistent with our empirical knowledge of plant behaviour. This means that the root:shoot ratio of a plant is *not* a result of an optimisation of the relative growth rate. The root:shoot ratio of a plant is therefore determined by other constraints.[9]

9.4   Nutrient use efficiency and the Piper-Steenbjerg effect

9.4.1   *Nutrient use efficiency*

Nutrient use efficiency (NUE) is a concept commonly used to characterise plant growth. The definition is simply the amount of biomass produced per unit of nutrient in the biomass, or the inverse of the nutrient concentration. There is, thus, a similarity between NUE and the nutrient productivity concept. The basic idea behind both concepts is that they should reveal the efficiency with which a plant makes use of the nutrients it has acquired. The difference between the concepts is that whereas nutrient productivity is dynamic, NUE is static; the units of the two are, for example, g dw $(g N)^{-1}d^{-1}$ and g dw $(g N)^{-1}$, respectively. Since plant growth is a dynamic process, the nutrient productivity should be closer to growth as such. Furthermore, it depends only on plant properties. The NUE, on the other hand, includes not only plant properties but integrates these over time and weighs them with the rate of supply of the nutrient. This makes NUE extremely sensitive to the temporal pattern of nutrient supply. We will provide some examples below.

The difference between the two concepts appears clearly when specific uptake patterns are considered. In Figure 8.10 we can see that plants growing at exponential rates but with different nitrogen productivities may have equal NUE (equal c_N, the vertical broken line). However, the plant with highest nitrogen productivity will be growing much more rapidly than the one with the lowest. Thus, in this case, NUE is no measure of the production potential of a plant. In Figure 8.10 it is only when nitrogen no longer is supplied at an exponential rate and nitrogen concentrations begin to drop that the most rapidly growing plant appears with a higher NUE.

One way of avoiding this problem with NUE is to find a way of defining the time during which a given quantity of a nutrient is operating in the plant.

Berendse & Aerts (1987) suggested that the mean residence time of a nutrient, t_r, should be used. Then, $\text{NUE} = P_N t_r$, but this equality is only useful for situations where a t_r can be defined.[10]

9.4.2   *The Piper-Steenbjerg effect*

The anomalous, negative, relation between plant growth and plant nutrient concentration that can occur with the nutrient use efficiency is also illustrated by the phenomenon called the Piper-Steenbjerg effect.[11] It may, however, again, be explained by the dynamic interaction between a nutrient-controlled plant growth and plant nutrient uptake.[12]

Consider therefore a plant that is extracting nutrients from a non-replenished medium according to

$$\frac{dN}{dt} = f_r V_d \frac{dW}{dt} \frac{N_V - \alpha(N(t) - N(0))}{V} \tag{9.13}$$

where f_r is the root fraction of the plant, V_d the volume that one unit of new root can deplete of nutrients,[13] N_V the initial amount of the nutrient in the medium, V the volume of the medium, and $0 \le \alpha \le 1$ a measure of the mobility of the nutrient in the root medium. Together with (8.1) and parameters characteristic for nitrogen-limited *Betula pendula*, Wikström (1994) calculated, among others, the relations between plant size and plant nutrient concentration shown in Figure 9.8. Plant species differ in their maximal capacity to take up nutrients,[14] we have therefore included this aspect in Figure 9.8.

However, small shifts in the temporal pattern of uptake result in drastically different relationships. For $c_{N,min} = 0$, and as long as $c_N < c_{N,opt}$, the system of equations (8.1) and (9.13) can be solved exactly to give

$$W(t) = W_0 + \frac{V}{\alpha f_r V_d} \ln\left(1 + \frac{N_0}{\dfrac{N_V}{\alpha} + N_0}\left(\exp\left\{ \frac{\alpha P_N f_r V_d}{V}\left(\frac{N_V}{\alpha} + N_0 \right)t \right\} - 1 \right) \right)$$

$$\tag{9.14}$$

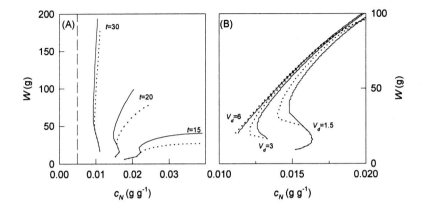

Figure 9.8 (A) Relation between plant mass and plant nitrogen concentration for plants starting with different N_V and with different harvesting dates. The plants are asymptotically approaching the broken line. $V_d = 1.5$. (B) Relations between plant mass and plant nitrogen concentration on day 20 for plants starting with different N_V and for three different V_d. Solid line: unlimited nitrogen uptake capacity. Dotted line: maximum relative uptake rate of nitrogen = r_m.

$$N(t) = \frac{N_0}{\dfrac{N_0}{\dfrac{N_V}{\alpha} + N_0} + \left(1 - \dfrac{N_0}{\dfrac{N_V}{\alpha} + N_0}\right) \exp\left\{-\dfrac{\alpha P_N f_r V_d}{V}\left(\dfrac{N_V}{\alpha} + N_0\right)t\right\}}$$

(9.15)

The relationship between plant mass and nitrogen concentration for (9.14) and (9.15) and with parameters identical to those in Figure 9.8 is shown in Figure 9.9A. In this case, plant weight increases with nitrogen concentration for low plant weights but decreases above a certain plant weight. This pattern is only a mirror image of that in Figure 9.8. What causes the difference between (9.13) and the numerical solutions derived by Wikström in Figure 9.8? The answer is found if we look at Figure 9.9B. Plants given sufficiently large amounts of nutrients in the medium will start by initially increasing their nutrient concentration. However, when the nutrient supply is exhausted, the nutrient concentration will start to drop rapidly and may eventually become lower than that of plants starting with an initially smaller supply. Of the two plants shown in Figure 9.9A, the one with the largest supply (largest N_V) will be the first to fall below the plant with the smallest N_V in concentration, Figure 9.9B. If we now introduce a ceiling for the plant nutrient concentration, $c_{n,opt}$, such that the plant with the largest N_V hits this ceiling, the decrease in its nutrient concentration is delayed and the order in which the plants with high nutrient supply drop to the concentrations of those with low supplies is reversed, Figure 9.8.

This analysis gives two other examples of how negative correlations between plant size and plant nutrient concentration can occur. In this case, it is caused by the interaction between growth and uptake; cf. Figure 8.10 where this negative correlation was caused by differences in P_N. It is therefore safe to conclude that plant nutrient concentration is not a reliable indicator of plant growth, except in cases of exponential growth when c_n is constant (8.4).

Problem 9.3.
Derive (9.14). (Simpler version, show that (9.14) satisfies (9.13) and (8.1).)

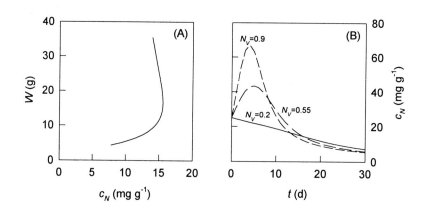

Figure 9.9 (A) Relation between plant mass and plant nitrogen concentration for plants starting with different N_V calculated from (9.14); conditions otherwise as in Figure 9.8. (B) Temporal development of nitrogen concentration for three plants starting with different N_V.

NOTES

[1] Ingestad & Ågren (1995).

[2] The sample is small but there seems to be a grouping of the species: conifers, deciduous trees, and the Lemna species. It might be profitable to speculate on the evolutionary significance of this interspecific variability in nitrogen productivity.

[3] Ågren (1985a) gives some examples of assumed fixed amounts and linear supply rates but it is not clear from the experimental set-up how accurately the assumptions about the uptake rate correspond to the actual conditions.

[4] Burns (1994).

[5] Ingestad & McDonald (1989) with birch and alder seedlings and Ingestad et al. (1994b) with birch. Earlier, Macdowall (1972) has shown that the relative growth rate at non-limiting nutrition follows a rectangular hyperbola of light intensity.

[6] Ågren & Ingestad (1987).

[7] Ågren & Ingestad (1987) expanded this discussion and also demonstrated the agreement between these predictions and experiments. Levin et al. (1989) modified the analysis to avoid the appearance of the maximum in the root:shoot curve.

[8] The weak effect of light on the root:shoot ratio is also observed in many experiments, e.g. Zelawski et al. (1985), Hunt & Nicholls (1986).

[9] This conclusion is, however, controversial as Hilbert (1990) argues from a somewhat different starting-point that the root:shoot ratio, indeed, can be construed as a result of optimising the relative growth rate. His result seems to come from the assumption of a "specific root activity" that is independent of the plant nutrient concentration. Inclusion of additional terms corresponding to, e.g. labile carbon compounds (Johnson & Thornley 1987), can add negative terms in (9.10) and avoid this problem. The addition of such factors seems, however, to lead to nutrient productivities that are inconsistent with observations (Ingestad & Ågren 1991).

[10] Grubb (1989) discusses further definitions and uses of NUE.

[11] Piper (1942), Steenbjerg (1951).

[12] Wikström (1994).

[13] A measure of root absorbing power, Nye & Tinker (1977).

[14] Mattson et al. (1991).

PART IV

THE ECOSYSTEM

When you've hit a real tough one, tried everything,
racked your brain and nothing works, and you know
that this time Nature has really decided to be difficult,
you say, "Okay, Nature, that's the end of the nice guy",
and you crank up the formal scientific method.

(Robert M. Pirsig *Zen and the art of motorcycle maintenance*)

10

Elements of an ecosystem theory

In this chapter we will look at the interaction between plants and soils in an ecosystem. We will start with an extremely general formulation and then proceed to analyse it by specifying its components with different models. These models are the descriptions of the plants and soils discussed in the previous chapters.

10.1 The general terrestrial ecosystem equation

We will now integrate the analyses and descriptions of the plant and the soil subsystems in the previous two sections into an ecosystem formulation. The representation of a terrestrial ecosystem we intend to use is one of a vegetation system linked to a soil system through their exchange of matter. In line with what we have developed earlier, we divide matter into carbon and nutrients (where the latter for the moment is only nitrogen). To make the vegetation component compatible with the soil system, we need to generalise its description. We will therefore not express the vegetation components in morphological terms but rather look at how the assimilated carbon and nutrients are distributed over the same kind of quality spectrum as we have studied within the soil, see further Section 12.2 below. Hence we define the plant carbon and plant nitrogen distributions, $\rho_{vC}(q,t)$ and $\rho_{vN}(q,t)$, respectively. Our view of an ecosystem can then be summarised as in Figure 10.1.

The formal representation of an ecosystem corresponding to Figure 10.1 is

$$\frac{\partial \vec{\rho}(q,t)}{\partial t} = \mathbf{E}\vec{\rho}(q,t) \qquad (10.1)$$

where $\vec{\rho}(q,t) = (\rho_{vC}(q,t), \rho_{vN}(q,t), \rho_{sC}(q,t), \rho_{sN}(q,t))$ is the ecosystem state vector and

$$\mathbf{E} = \begin{pmatrix} -\mu_v(q) & \int dq'\, P_N(q,q') & 0 & 0 \\ -r_{0N}(q)\mu_v(q) & 0 & -A_N(q)\int dq'\, f_N(q')D(q,q')u(q') & A_N(q)f_C\dfrac{u(q)}{e(q)} \\ \mu_v(q) & 0 & -f_C\dfrac{u(q)}{e(q)}+f_C\int dq'\, D(q,q')u(q') & 0 \\ r_{0N}(q)\mu_v(q) & 0 & \int dq'\, f_N(q')D(q,q')u(q') & -f_C\dfrac{u(q)}{e(q)} \end{pmatrix}$$

$$(10.2)$$

is the ecosystem operator. $P_N(q,q')$ is a generalisation of the nutrient productivity describing the amount of carbon of quality q that will be produced per unit time by nutrients bound at quality q'. Since the nutrients made available to the vegetation appear in inorganic form they are not associated with any quality. We need therefore an allocation factor, $A_N(q)$, that distributes the nutrients taken up over the quality spectrum.

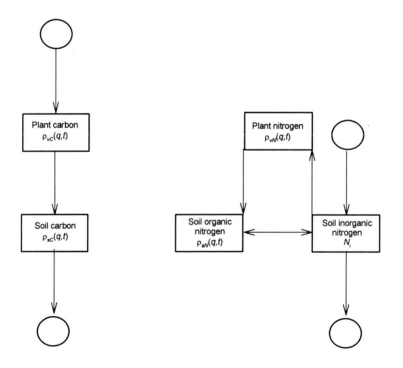

Figure 10.1 Schematic representation of a terrestrial ecosystem. The circles stand for external sinks and sources.

Equation (10.1) is very general and we are only able to analyse it by applying specific models. We will do this in the following sections of this chapter. There is, however, one point to be made. This description of an ecosystem is basically a positive feedback system. For example, increased plant nitrogen increases plant production, which increases litter fall, which increases soil carbon, which increases soil carbon mineralisation, which increases soil nitrogen mineralisation, which increases plant nitrogen uptake, and hence plant nitrogen. That the system does not run amok depends on two factors. Firstly, the total amount of nitrogen in the ecosystem is limited. Constraints from another, higher level of organisation therefore place limits on what the system can do. Secondly, there are also internal negative feedbacks that will tend to keep the system in check. For example, nitrogen productivity declines with increasing plant carbon, equation (9.5).

Equation (10.1) is, of course, not a complete statement of all processes in an ecosystem but portrays a certain level of representation with respect to both temporal and spatial resolution. We have, for example, excluded the soil inorganic nutrient pool from our state variables with the idea that its turnover is rapid compared with the organic components and it should therefore be in equilibrium with respect to the surroundings. We have also excluded any explicit representation of other living organisms than the vegetation. The decomposer community responsible for the decomposition of the organic material is included implicitly, see Section 4.2, but other organisms, such as herbivores, are neglected.

10.2   Ecosystem stoichiometry

Plants and decomposers have limited flexibility in their elemental ratios, with plants generally having lower or even much lower nutrient:carbon ratios than the decomposers. These limitations also impose limits on how nutrient:carbon ratios of the ecosystem as a whole may vary. Under the heading of ecosystem stoichiometry[1] we intend to explore these questions. Let us start with a simple, closed system in which the total nitrogen content, N_T, remains constant; the inorganic nitrogen is small so we will set it to zero. The vegetation is represented by only one carbon, C_v, and one nitrogen, N_v, component and similarly for the soil, C_s and N_s, respectively. We will also treat the soil as a homogeneous substrate; adding heterogeneity would not make any difference in this case.[2] The ecosystem is then described by the following four dynamic equations

$$\frac{dC_v}{dt} = (a - bC_v)N_v - \mu_v C_v \tag{10.3}$$

$$\frac{dN_s}{dt} = -k_N N_s + r_c k_N C_s + \mu_{vN} N_v \tag{10.4a}$$

$$\frac{dN_s}{dt} = -k_N N_s + r_c k_N C_s + r_0 \mu_v C_v \tag{10.4b}$$

$$\frac{dC_s}{dt} = -k_C C_s + \mu_v C_v \tag{10.5}$$

$$N_v + N_s = N_T \tag{10.6}$$

where $k_N = f_C u / e_0$ (7.1), $k_C = (1 - e_0) f_C u / e_0$ (3.13), and $r_c = e_0 f_N / f_C$ is given by (3.5). We have two equations describing the dynamics of N_s because we want to explore two alternative mechanisms for nitrogen losses from the vegetation. The empirical evidence is ambiguous on this point. Equation (10.4a) assumes that there is a constant specific loss rate of nitrogen whereas (10.4b) assumes that there is a constant nitrogen concentration in the litter.

At steady state we have from (10.3)

$$r_v = \frac{N_v}{C_v} = \frac{\mu_v}{a - bC_v} \approx \frac{\mu_v}{a} \quad \text{for } b \to 0 \tag{10.7}$$

and from (10.4b) and (10.5)

$$r_s = \frac{N_s}{C_s} = r_c + \frac{k_C}{k_N} r_0 = r_c + (1 - e_0) r_0 \tag{10.8}$$

These two ratios can then be combined to give an ecosystem nitrogen:carbon ratio

$$r_e = \frac{N_v + N_s}{C_v + C_s} = \frac{k_C}{\mu_v + k} r_v + \frac{\mu_v}{\mu_v + k_C} r_s \tag{10.9}$$

The ecosystem nitrogen:carbon ratio comes out simply as a weighted average of the corresponding nitrogen:carbon ratios of the plant and soil, respectively, and the weighting factors are the decay rates of the plant and soil. The last term in (10.9) is the dominating term as $r_v < r_s$ and $k_C < \mu_v$.

Hence, plant mortality relative to soil carbon decomposition rate and soil nitrogen:carbon ratio become the most important factors.

Problem 10.1
Derive (10.9).

We can also calculate how carbon and nitrogen is partitioned between the vegetation and the soil.

$$\frac{C_s}{C_v} = \frac{\mu_v}{k_C}$$
(10.10)

and

$$\frac{N_s}{N_v} = \frac{r_0 + r_c(1 - e_0)}{k_C / a}\left(1 - \frac{C_v}{a / b}\right) = \frac{\mu_v}{k_C}\frac{r_s}{r_v}$$
(10.11)

The relative carbon distribution should thus be constant whereas the nitrogen distribution depends on the fertility of the site, C_v increases with increasing nitrogen availability, cf. Figure 11.1 below.

Problem 10.2
Derive (10.11).

The results derived so far in this section are all based on the assumption that the vegetation can be treated as one homogeneous mass. Let us remove this restriction and create plants that consist of leaves (subscript *l*) and woody biomass (subscript *w*). We then need to add four additional equations to the system already defined.

We assume that nitrogen is incorporated in the woody biomass with a concentration of r_{0w} and this is also the concentration in the woody litter fall. The growth of the woody biomass is assumed to be proportional to the growth of the leaf biomass, with the proportionality factor g. We can then add two new equations

$$N_{vw} = r_{0w}C_{vw}$$
(10.12)

and

$$\frac{dC_{vw}}{dt} = g(a - bC_v)N_{vl} - \mu_{vw}C_{vw}$$
(10.13)

The equation for soil carbon from woody biomass is similar to (10.3) and there will be two equations for soil nitrogen similar to (10.4). For simplicity, we will assume the same r_c for nitrogen associated with leaves and wood.

The results obtained with this more complicated system are quite similar to (10.9)-(10.11) but with g as a correction factor. Thus,

$$\frac{C_s}{C_v} = \frac{C_{sl} + C_{sw}}{C_{vl} + C_{vw}} = \frac{\mu_{vl}}{k_{Cl}} \frac{1 + g\dfrac{k_l}{k_w}}{1 + g\dfrac{\mu_l}{\mu_w}} \tag{10.14}$$

$$\frac{N_s}{N_v} = \frac{N_{sl} + N_{sw}}{N_{vl} + N_{vw}} = \left[\frac{r_c + (1-e_0)r_{0l}}{k_{Cl}} + g\frac{r_c + (1-e_0)r_{0w}}{k_{Cw}}\right] \frac{(a - bC_{vl})}{1 + g\dfrac{r_{0w}}{\mu_{vw}}(a - bC_{vl})} \tag{10.15}$$

and for $b = 0$

$$r_v = \frac{N_{vl} + N_{vw}}{C_{vl} + C_{vw}} = \frac{\mu_{vl}}{a} \frac{1 + \dfrac{ag}{\mu_{vw}}r_{0w}}{1 + g\dfrac{\mu_{vl}}{\mu_{vw}}} \tag{10.16}$$

$$r_s = \frac{N_{sl} + N_{sw}}{C_{sl} + C_{sw}} = \frac{r_c + (1-e_0)r_{0l} + g\dfrac{k_{Cl}}{k_{Cw}}\left[r_c + (1-e_0)r_{0w}\right]}{1 + g\dfrac{k_{Cl}}{k_{Cw}}} \tag{10.17}$$

An equation for the ecosystem N:C ratio, r_e, can also be derived but its complexity makes it uninteresting. It is also evident that more complex systems are mostly beyond the scope of analytical derivations.

10.3   Comparison with other approaches

The understanding of ecosystem functioning is still in its infancy. The approach we have indicated above is only one of a number of partially competing, partially complementing, approaches. A key problem seems to be which questions are supposed to be answered, with a major dividing line between those that pertain to derive the most favourable plant traits in a given environment (soil and climate) and those that take the plant properties for granted and investigate its consequences. Grime's (1979) three strategies and Tilman's (1988) allocation models should be included within the former groups, whereas some models to be discussed further below belong to the latter group. Field et al. (1992, see also Shaver & Aber 1996) suggest a resource-based theory of ecosystem functioning in which both plants and soil respond to changes in environment. Since we are assuming constant plant properties, except that allocation to vegetative organs can change with nutrient availability (section 9.3), we will be answering different questions to those posed by Grime or Tilman.

Our work is distinct from that of Field et al. (1992) in two respects. We set a much narrower scope for our investigations by letting a number of processes be represented only implicitly. Secondly, we only insist on a mathematical formulation to obtain the advantages discussed in Chapter 1. We can not yet see to what extent there are qualitative differences in the predictions from the approaches; quantitative comparisons are not possible.

10.3.1   *CENTURY, GEM, G'DAY*

There are numerous ecosystem models that closely resemble our Figure 10.1; for a review see Ågren et al. (1991). Some of the most similar are CENTURY (Parton et al. 1987), GEM (Rastetter et al. 1991) and G'DAY (Comins & McMurtrie 1993). All these models follow the same conceptual format but with the plant and soil components somewhat differently aggregated. In all of them, plant growth is based on the balance between photosynthesis and respiration although the development of the plant component in G'DAY relative to its predecessor BIOMASS (McMurtrie et al. 1992) has led to a formulation that looks very similar to the nitrogen productivity. The soil descriptions are also similar (G'DAY has actually taken its soil component from CENTURY) with a partitioning of litter and soil organic matter into a small number of discrete fractions with characteristic turnover rates.

10.3.2   *FORET, LINKAGES*

A different category of models is made up of those where species replacement plays a dominant role. Prominent models are FORET (Shugart 1984) and LINKAGES (Pastor & Post 1986). These models are characterised by fairly simple phenomenological descriptions of plant (tree) growth but where competitive interference between individuals is important. In most applications of these models, nutrient availability is only expressed through a relative growth reduction when the potential growth exceeds the available amount of nutrients as estimated in the decomposition routines.

10.3.3   *MEL*

In order to test the importance of the flexibility of an ecosystem to adjust its resource utilisation, Rastetter et al. (1997, see also Rastetter & Shaver 1992) developed a highly aggregated model of a two-element cycle with vegetation and soils. This model, MEL, demonstrates the large effects that different adjustment mechanisms have at different time scales. It also points out the need to know the degree to which a system is closed, i.e. the need to know the flux of an element (e.g. nitrogen) through the system relative to the internal cycling.

NOTES

[1] Redfield (1958), Rastetter et al. (1992).
[2] In formal terms, the carbon and nitrogen distributions are modelled by delta functions.

11

Ecosystems - applications

In this chapter we take up some applications where interactions between plants and soil are important. Most of these applications go to a fairly high degree of detail in their descriptions of the ecosystems and here we forsake generality for representativeness. As a consequence, very few results can be derived analytically and only numerical soultions are available.

11.1 Ecosystems and global change

As a first application, let us use equations (10.3)-(10.6) to investigate potential changes in ecosystem functioning from a change in climate and to investigate which factors impose constraints on the ecosystem. Let us only consider how a climatic change may affect the steady state of the ecosystem. As long as we restrict ourselves to steady state situations we can use this simple description of the ecosystem. We will explore the two possibilities for nitrogen litter fall (10.4a) and (10.4b) in parallel. At steady state, the litter production equals primary production. We can then express primary production, P, in the ecosystem as

$$P = \mu_v C_v \qquad (11.1)$$

At steady state equation (10.3) can easily be rewritten to give P in terms of N_v [1]

$$P = \frac{aN_v}{1 + bN_v/\mu_v} \qquad (11.2)$$

This is the constraint from internal plant properties. We get a second constraint on the production from the nitrogen cycle when we combine (10.4), (10.5), and (10.6) to another expression of P in terms of N_v

$$P = \frac{k_C}{r_c}(N_T - \frac{\mu_{vN} + k_N}{k_N} N_v) \tag{11.3a}$$

$$P = \frac{k_C}{r_c + (1 - e_0)r_0}(N_T - N_v) \tag{11.3b}$$

The somewhat surprising result in (11.3) that maximum plant production occurs when all nitrogen is allocated to the soil is because, in order to derive (11.3), no nitrogen is required for plant growth. Instead, maximum plant growth occurs when C_s is as large as possible, (10.5). But carbon and nitrogen are coupled in the soil and the maximum C_s obtains when as much nitrogen as possible is allocated to the soil. The intersection between the two production lines is then the actual production. We have illustrated the consequences of these equations in Figure 11.1 where we assume that a climatic change increases plant production through its effect on b (b decreases by 10%) and the specific decomposition rate, k, increases by 10%.

Under the assumptions made, the effects on plant production alone are very small, about 1%. The major effect on the production rates comes from increased nitrogen availability, which boosts production by 15%. The combined effect of these two factors together is, however, larger than the sum of the effects of the two factors separately; the production increases by 17%, which is a small but nonetheless a synergetic effect. The increased production leads also to a larger store of carbon in the ecosystem, but the effects are rather small as the increase in plant carbon is off-set by the decrease in soil carbon. The largest change, 4%, obtains when both production and decomposition rates are changed. Because of the rather shallow slope of the nitrogen cycling lines in Figure 11.1, changes in only plant production properties are not efficient in changing production levels or carbon stores. More efficient changes obtain when nitrogen can be shifted between ecosystem components. It is, therefore, of interest to look more closely at the relations between carbon and nitrogen in the different ecosystem components. One should be observant of the fact that there are factors in a climatic change that might lead to a decrease in production. If an increased CO_2 concentration leads to litters of lower quality, both in terms of q_0 and r_0, one might even get a situation where the constraints from the nitrogen cycle (the straight lines in Figure 11.1) fall below the lines for unaltered climate.

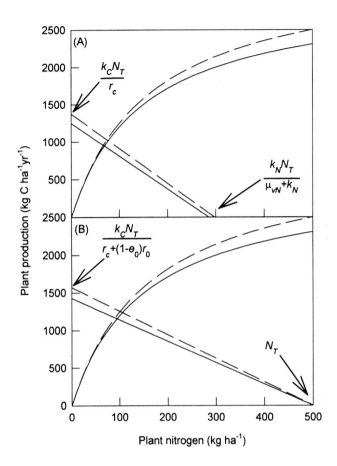

Figure 11.1 Relation between steady state plant production and plant nitrogen content. The solid lines represent equations (11.2) and (11.3a or b) with parameter values typical for a *Pinus sylvestris* canopy. The broken lines represents the same equations but with parameters changed to represent the effects of a climatic change.

Problem 11.1

Which parameter in the system of equations (10.3)-(10.6) is most efficient in changing the ecosystem carbon store?

11.2   Nitrogen saturation

In the previous section, we considered how elements were distributed within an ecosystem in relative terms. We should now ask ourselves, what determines the amounts of elements that can be stored in an ecosystem? In particular, we are interested in how much nitrogen an ecosystem can accumulate.[2] Could some of these ecosystems accumulate more, or have they reached a point where additional inputs would only result in losses? One of the reasons why we need to ask this question is, of course, the high rates of nitrogen depositions in large forest areas in north-western Europe and northern America (some tens of kg N $ha^{-1}yr^{-1}$).

We consider an ecosystem to be nitrogen-saturated when nitrogen losses from the system approximate or exceed the inputs of nitrogen. Since nitrogen normally is a scarce element in terrestrial ecosystems, the organisms in such systems are generally conservative with any nitrogen that may be available. Losses are therefore a sign of excess. The balance between inputs and losses should be calculated over a reasonably long time whereby temporary disturbances are cancelled out. The clear-cutting of a forest is often accompanied by leaching of nitrate, but after regrowth of the vegetation the losses disappear. In the short-term perspective, this may be interpreted as nitrogen saturation, but it is the long-term nitrogen saturation, e.g. where substantial nitrate leaching would occur over the whole rotation period of the forest, that is our concern.

A system can become saturated by two different mechanisms. On one hand, the input rate of nitrogen may be larger than the immediate uptake capacity of the system. In fertilisation trials in forests where hundreds of kg N ha^{-1} are commonly applied as a single dose, 60 to 70% of the applied fertiliser can be recovered in the system[3] demonstrating a high uptake capacity (See Chapter 7). Saturation of the uptake capacity is, therefore, something that should be rare and we will not consider it further. The other possibility of saturating the system is by adding, slowly or rapidly, such large, cumulated amounts of nitrogen that the internal cycling becomes saturated. The point where saturation should show up is at the interface between the mineralisation of soil nitrogen and plant uptake of inorganic nitrogen. Saturation is, therefore, not associated with the statics of the system but with its dynamics. One should finally note that high deposition of nitrogen during a dormant season, but which becomes washed out before the growing season starts again, should not be considered as nitrogen saturation as this transfer entirely bypasses the functioning of the system.

Since the largest and most dynamic store of nitrogen within the vegetation subsystem is the foliage, we will represent it with (10.3). At steady state we then have the following relation between foliage nitrogen concentration, foliage carbon and foliage content of nitrogen

$$C_v = \frac{ac_N - \mu_v}{bc_N} \tag{11.4}$$

and

$$N_v = \frac{ac_N - \mu_v}{b} \tag{11.5}$$

Let us describe the nitrogen losses in litter fall with (10.4b). The value of r_0 varies greatly between species and litter types. When Johansson (1995) investigated nitrogen concentrations in needle litters of Scots pine and Norway spruce from sites ranging from northern to southern Sweden and covering a wide range of site indices, she found that the nitrogen concentration in eight Scots pine litters varied between 0.38 and 0.43% of dry weight and in ten Norway spruce litters between 0.45 and 0.63%, which is consistent with (10.4b). Hunter et al. (1985), on the other hand, found a clear positive correlation between nitrogen concentration in live needles and needle litters corresponding to (10.4a). We prefer, however, to use (10.4.b).

The rate of nitrogen returned to the soil is $r_0\mu_vC_v$. At steady state, this is also the required rate of uptake. When the foliage nitrogen concentration is at its maximum value we get the maximum nitrogen flux density (nitrogen mineralisation plus deposition) that the plant system can utilise for its foliage. When the nitrogen flux density exceeds this level the plant system can no longer follow; an excess has been created. This is the indication that the plant system has reached saturation.

Consider now a soil that is receiving a constant rate of litter input, $I_0(= \mu_vC_v)$ and let us follow Ågren & Bosatta (1987) and use parameterisation I with $e_0 = 0.05$, $q_0/\eta_{10} = 1.12$, $\beta = 3$, and $1/(e_1\eta_{11}) = 6$. This corresponds to a long-term accumulation of around 1% for coarse woody litter and a negligible accumulation for other litter types. From (4.45) and (4.48) we then get

$$N_s(t) \approx \frac{f_N}{f_C} I_0 e^{\frac{q_0}{\eta_{10}}} \left[\frac{e_0}{e_0 + e_1 q_0} \right]^{\frac{1}{\eta_{10} e_1}} t \qquad (11.6)$$

showing that there will be a steady state linear increase in nitrogen in the system. This increase has to be sustained by external sources. The net mineralisation rate, m, can likewise be calculated

$$m(t) = I_0 r_0 \left\{ 1 - \left[1 + \frac{f_N / f_C}{r_0} \left(e^{\frac{q_0 - q_t}{\eta_{10}}} - 1 \right) \right] \left[\frac{e_0 + e_1 q_t}{e_0 + e_1 q_0} \right]^{\frac{1}{\eta_{10} e_1}} \right\} \qquad (11.7)$$

The first term within the braces, 1, represents equilibrium with the plant system, and the remaining terms the long-term immobilisation of nitrogen in the system. This term is therefore the ultimate rate at which the soil can absorb nitrogen depositions. Inserting our parameter values shows that even with as high a value of f_N/f_C as 0.2, the accumulation rate will not reach higher rates than 0.5% of the leaf nitrogen fall. Thus, since the leaf nitrogen fall[4] can amount to 40 to 50 kg N ha^{-1}yr^{-1}, the possible accumulation rate of nitrogen in the soil is 0.2 kg N ha^{-1}yr^{-1}. The corresponding accumulation rate for the woody litter is still smaller.

The rate of mineralisation can be expressed also as a fraction of the amount of nitrogen in the soil, i.e. $m(t)/N_s(t)$. This relative rate of mineralisation is shown in Figure 11.2 for soil organic matter derived from litter of needles or coarse woody material. There is an approximately 50-fold difference between these two types of litters. The importance of the age of the soil is also clear, where the changes are particularly emphasized at young ages. The rapid increase in mineralisation of woody materials is a reflection of a switch from a stage of immobilisation (negative values) to net mineralisation.

Equation (11.7) shows that the rate of net mineralisation from a given litter fraction is always less than the influx of nitrogen to the same litter fraction, when the litter production rate is constant. As a consequence, under quasi-steady state conditions, the soil can never deliver more nitrogen than the plant system can take up. The simple conclusion is that the soil system will saturate before the plant system.

Let us calculate how rapidly the soil system will become saturated under different deposition scenarios. For simplicity, assume that the soil is old, such

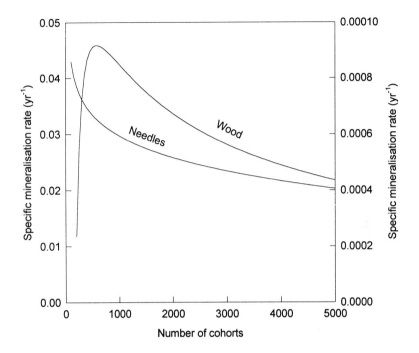

Figure 11.2 Net mineralisation rate relative to soil nitrogen amount as a function of the number of litter cohorts having been added to the soil (soil age). Left scale for needle litter and right scale for woody litter (from Ågren & Bosatta 1988).

that the quality of the oldest litter cohort can be set to zero. The accumulation of nitrogen with increasing soil carbon is small (11.6). A larger potential for storing nitrogen is an increasing decomposer nitrogen concentration. Let us consider how f_N/f_C must increase in order to absorb a continuous deposition of U kg N ha^{-1}yr^{-1}. Starting from a steady state situation at $t = 0$ requires an f_N/f_C at time t_p of

$$\frac{f_N(t_p)}{f_C} = \frac{f_N(0)}{f_C} + k\frac{U}{r_0 I_0} r_0 t_p \qquad (11.8)$$

where k is a numerical constant, the value of which depends on the age of the soil and the initial litter quality. Thus, for a given litter, the rate of change in f_N/f_C will depend upon the deposition relative to the litter production rate ($U/r_0 I_0$) and the litter nitrogen concentration (r_0). The required increase in

f_N/f_C is shown in Figure 11.3 for two soil ages (500 and 5000 years, respectively, giving $k = 0.331$ and 0.188) and four deposition rates (25, 50, 75, and 100% of litter fall) assuming the soil to be built up of needle litter. If f_N/f_C can be increased by 0.1, the soil should be saturated within 25 to 50 years under the most heavy deposition scenarios. However, if the soil organic matter is derived from woody material, the rate of increase in f_N/f_C is an order of magnitude slower ($k = 0.0781$ and 0.0189 for soil of ages 500 and 5000 years, respectively). Another consequence is that, even if r_0 were to increase in response to nitrogen deposition, the approach to saturation as described in Figure 11.3 would not differ, because of the way r_0 appears in (11.8). An increased productivity would, on the other hand, slow down the approach to saturation.

11.3   Short rotation forestry

It is hoped that short rotation forestry with intensive management can provide large amounts of biomass for use as an energy source. If, by high

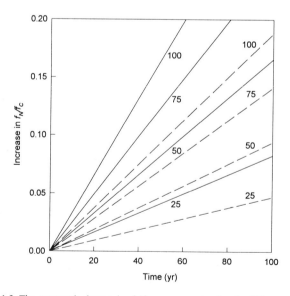

Figure 11.3 The temporal change in f_N/f_C necessary to absorb different rates of nitrogen deposition by a soil formed from needle litter. The values next to the lines are deposition rates in per cent of nitrogen litter fall, soil ages 500 years (solid lines) and 5000 years (broken lines) (from Ågren & Bosatta 1988).

production, we mean 20 000 kg dw ha^{-1}yr^{-1} of harvestable wood (*S*), we will also require simultaneous production of leaf biomass (*L*) of the order of 6 500 kg ha^{-1} and of fine roots (*R*) of 6 250 kg ha^{-1}. With a nitrogen concentration of 0.04 kg N (kg dw)$^{-1}$ in leaves (n_{opt}) and fine roots (n_R) and 0.006 in wood (n_S), respectively, the annual turnover of nitrogen is 630 kg ha^{-1}, out of which 120 has to come from external sources to replace what is harvested with the wood. The remaining 510 kg ha^{-1} should come from mineralisation of old leaves and fine roots. To maintain such a high rate of nitrogen turnover without losses requires high precision in the timing of supply and uptake. We shall now consider this problem in greater detail.[5]

Let us, therefore, describe crop growth for the period 1 June (day 1) to 15 September (day 105) in the following way (slightly modifying the models of the previous sections)

$$\frac{dL}{dt} = (an_{opt} - bLn_{opt})L \tag{11.9}$$

$$\frac{dS}{dt} = p_S L \tag{11.10}$$

$$\frac{dR}{dt} = p_R L \tag{11.11}$$

$$U = n_{opt}\frac{dL}{dt} + n_s\frac{dS}{dt} + n_R\frac{dR}{dt} \tag{11.12}$$

where p_S (= 0.052 d^{-1}) and p_R (= 0.016d^{-1}) are parameters linking growth of stems and roots to leaf biomass and *U* the uptake rate of nitrogen. Between 15 September and 15 October (day 135) there is no new production of leaves, instead the leaf biomass decreases linearly to zero. The parameters *a* and *b* have been given the values 1.34 kg dw ha^{-1}yr^{-1} and 0.00183 ha (kg N)$^{-1}$d^{-1}, respectively.

We also need to modify the description of nitrogen mineralisation to be able to follow the within-year variations in mineralisation rates. Let us assume that the major driving force behind the variation in the rate is the climatically driven changes in decomposer growth rate. The variations in soil temperature can be approximated with a cosine function, and if we let the decomposer growth rate follow this variation we get

$$u(q,t) = u_1(q)u_2(t) \tag{11.13}$$

with $u_1(q)$ given by (4.42) and

$$u_2(t) = [1 + \cos(\omega t + \varphi)]/2 \tag{11.14}$$

With this model for the decomposer growth rate and parameterisation I with $e_0 = 0$ and other parameter values as for Scots pine needles (Table 5.1), we can calculate the relation between q and t (7.25)

$$\frac{q(t)}{q_0} = \frac{1}{\left[1 + f_C \eta_{10}(\beta - 1)u_0 q_0^{\beta - 1}(t + \dfrac{\sin(\omega t + \varphi)}{\omega})/2\right]^{\frac{1}{\beta - 1}}} \tag{11.15}$$

By varying φ we can change the time when nitrogen mineralisation starts relative to plant growth ($\varphi = 0$ for simultaneous start), and by varying ω we can adjust the length of the period of mineralisation.

Figure 11.4 shows the consequences of shifts in the start of nitrogen mineralisation. A shift already as small as 15 d has drastic effects. If nitrogen mineralisation starts ahead of uptake, a small initial excess (less than 10 kg ha^{-1}) is created followed by a deficit from approximately day 60 of the growing season. This deficit reaches a maximum of around 140 kg ha^{-1}, which is essentially what is required for wood growth. On the other hand, if nitrogen mineralisation is delayed vis-à-vis growth, the maximum increases rapidly, attaining as much as 270 kg ha^{-1} for a 15-day delay. Under such conditions, it would be necessary to store as much as 150 kg N ha^{-1} (270-120) as buffer from one year to the next. This is a considerable amount and leaching losses are likely to occur.

11.4   Other applications

All the analyses have so far been made on systems, that have been simplified to the extent that analytical solutions have been possible. These simplifications have, of course, been made at the expense of the detail in the representation of specific ecosystems. Emphasis has instead been placed on qualitative behaviour in the hope that the principal modes of the behaviour of the systems should be clearer. We have, however, also worked with more complex models that include more details. In these applications, only numerical results are available. Leaving steady state conditions also often

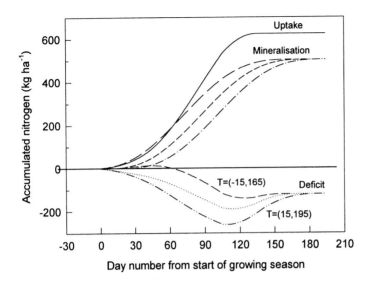

Figure 11.4 Accumulated uptake of nitrogen by the vegetation (solid line), accumulated nitrogen mineralisation (three upper broken lines), and deficit in nitrogen supply (three lower broken lines) during a growing season when nitrogen mineralisation starts and ends at different times. For example, T = (15,195) means that nitrogen mineralisation starts 15 days after of plant growth and ends on day 195 relative to the start of plant growth (from Ågren 1989).

necessitates recourse to numerical methods. We have developed a simulation model, Q,[6] that includes an extensive description of the linked carbon-nitrogen cycles of forests dominated by an evergreen species (to allow us to neglect within-year variations) and all other plant species combined into a generic "grass". The formulation of the model follows the lines outlined earlier in this chapter but the soil carbon and nitrogen dynamics are described with equations for heterogeneous substrates. Since litter inputs will vary in time, the simple calculations of Chapter 5 with constant rates of inputs are no longer possible. Instead we need to follow each litter cohort individually during the entire simulation. The model is schematically presented in Figure 11.5.

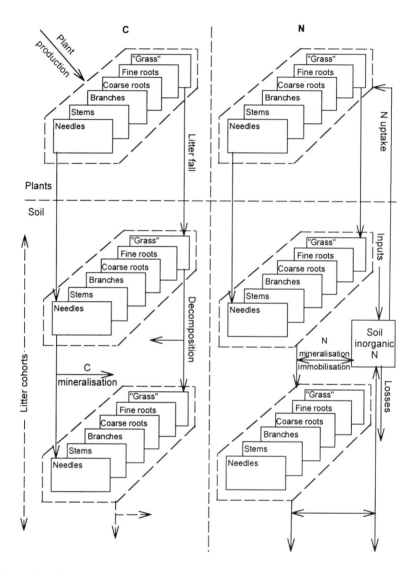

Figure 11.5 Schematic representation of the Q-model showing flows and state variables. The needles are further subdivided into age classes (from Ryan et al. 1996a).

11.4.1 *Global change - revisited*

Using the Q model we have analysed consequences of climatic change over a rotation period for two different coniferous forests; a *Pinus sylvestris* forest in Sweden (the SWECON site), and a *Pinus radiata* forest in Australia (the BFG site).[7] The main purpose of the study was to investigate the consequences of climatic change on carbon reservoirs. The sites were therefore compared with four different scenarios of climate change: (i) the climate remains unaltered; (ii) the temperature increases by +4 °C; (iii) the atmospheric CO_2 concentration increases to 700 ppm; and (iv) the combination of temperature and CO_2 increases. The results point to an increase in net primary production, Figure 11.6, but the effects of an increase in temperature and CO_2 are not additive, and instead it seems that with increased temperature a maximum in production is reached with little additional effect of CO_2. The effects of CO_2 alone are more complicated. At the BFG site, CO_2 alone increases production almost as much as temperature but at the SWECON site the increased CO_2 initially increases the production almost as much as temperature alone, but in the later phase of the rotation period the effects have levelled off and the production rate is almost at par with the reference situation.

To interpret what happens, we need to look at further details of the results. In particular we should have a look at the development of needle biomass, Figure 11.7. The small response at the BFG site is consistent with the small effects on needle biomass, indicating that this stand is already operating close to its potential maximum rate. The interactions between the carbon and nitrogen cycles are also important. If Figure 11.1 was redrawn for a system with larger a and b but equal a/b and N_T it is clear that the response would be smaller. At the SWECON site, on the other hand, the needle biomass can develop much more rapidly under the more favourable conditions. However, a consequence of this development is that a large quantity of the system nitrogen becomes locked up in the woody components of the trees at an earlier stage of stand development with the subsequent slowing down of needle biomass development. When the temperature is increased, there is not only an increase in plant production but also in nitrogen mineralisation, which may compensate for the nitrogen withdrawn in the woody biomass. The absolute numbers should be considered with a measure of caution as the estimates of response surfaces for all parameters to the climatic variables are not fully

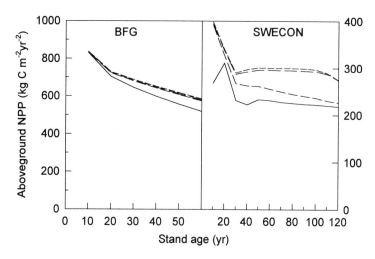

Figure 11.6 The effects of climate change on aboveground net primary production at the BFG site and the SWECON site. The solid line represents the unaltered climate and the three broken lines the three different climate change scenarios. The lowest of these three lines represents CO_2 alone, next comes temperature alone, and at the top the two factors combined (modified from Ryan et al. 1996b).

satisfactory. The importance of the feedback loops between the carbon and nitrogen cycles for interpretation and understanding of climatic change effects are, however, clear.

11.4.2   *Forest nutrient budgets*

A major substitute for fossil fuels is biomass. This biomass can be produced in specialised production systems, e.g. in short rotation forestry as described in Section 11.3, or by more intensive harvesting in conventionally managed forests. A potential problem with more intensive harvesting is that the increased removal of biomass will remove so much extra nutrients that future production will suffer. We have used the Q model to study different manners of extracting more biomass in order to see which factors are most likely to impair sustainable production.[8]

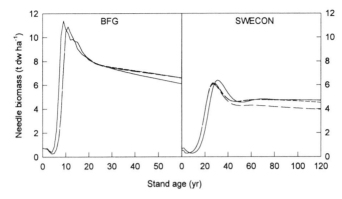

Figure 11.7 The effects of climate change on needle biomass development. Symbols as in Figure 11.6.

Differences in site productivity are one of the factors likely to determine the sensitivity of a stand to management. We have therefore considered three site classes, G16, G24, and G32, where G stands for gran (the Swedish word for Norway spruce, *Picea abies*) and the number following the G is the mean height of the dominating trees at age 100 yr. The three site classes were chosen to represent a range from a poor site over a medium site to the best site found under Swedish climatic conditions. The model produces stem growth rates that compare well with rates listed in conventional yield tables, Figure 11.8. The major differences occur in the earliest phase of stand development and are due to problems of correctly estimating the fraction of the soil volume used by the seedlings. Nitrogen mineralisation is taking place in the entire soil volume but the root systems of the trees can initially only exploit a small fraction of this soil volume.

The effects of repeated harvests of varying intensity differ markedly between the three site classes. Figure 11.9 shows the development of three rotations in a G16 stand with two levels of removal of branches and needles (50 and 100%) at the final harvest compared with conventional harvesting where only stem biomass is removed both at thinnings and final harvests. The effects in this case are drastic, with the harvest at the end of the third rotation being only half of that at the first in the most intensive harvesting programme. The more productive sites respond much less and the decrease in harvest is only about 15% with the same comparison. The larger decrease in harvest on the low site class is not only a consequence of removing nitrogen in the harvested material but there is also an effect on the nitrogen conservation in the phase of stand development immediately following the harvest at the end of a rotation period. Woody material (branches) left on the ground after a harvest will immobilise nitrogen mineralised from needles. If the plants at that moment have a low potential to take up nitrogen, this is a mechanism that will prevent leaching losses and conserve nitrogen for the future. This kind of conservation is more important in low productive stands where the slow development of the trees implies a longer time before the entire soil volume can be exploited and leaching prevented. Hence, leaching losses in low-productive stands are relatively more important in such places. For a similar reason, removal of branches and needles in thinnings has been found to have a larger effect on future production than removals at final harvests. The removal of nitrogen at thinnings is removal of nitrogen that otherwise would have rapidly been returned to normal circulation, whereas removal of nitrogen at final harvest is largely removal of nitrogen that anyhow would have leached out of the system.

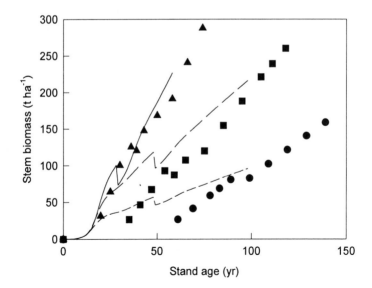

Figure 11.8 Predictions of stem biomass development for three yield classes of Norway spruce (G16, G24, and G36) according to the model (lines) and according to yield tables (symbols). Thinnings at 30 or 50 years are also included (from Rolff & Ågren 199X).

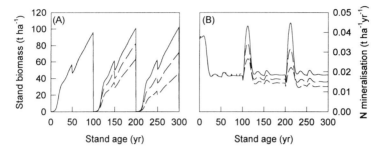

Figure 11.9 Aboveground biomass (A) and nitrogen mineralisation rate (B) in a G16 stand with only stem harvesting (solid line) and 50% or 100% removal of branches and needles (broken lines) at final harvests (from Rolff & Ågren 199X).

11.4.3 *Acid depositions to forest ecosystems*

Changes in deposition rates of elements to forest ecosystems will change the soil chemistry. Some of these changes are directly harmful, e.g. acids that decrease soil pH and thereby release toxic elements. Others are less obvious, nitrogen deposition on one hand increases the availability of a limiting element but on the other hand contributes to acidification. One must consider what happens when nitrogen is no longer limiting due to deposition. We have investigated what happens when, instead, magnesium becomes the limiting element with a model akin to the Q model.[9] In this model, the cycle of two elements (N and Mg) and the detailed carbon and nitrogen cycles in the soil are replaced by the homogeneous equations and a soil chemical module to take into account the interaction between N and Mg with other ions (SO_4^{2-}, Ca^{2+}, H^+, HCO_3^-, and $Al(OH)_x$-species).

Two scenarios for the development of a Norway spruce (*Picea abies*) forest in central Europe were analysed. In one scenario, the deposition of elements continues from 1990 and onwards at the same level (no reduction scenario). In the other scenario, deposition of N and S decreases by 7.5% per year between the years 1990 and 2000 (improved environment scenario). Initially, the forest is nitrogen-limited. However, because the external supply of N and Mg is not in balance with the requirements of the trees, and because Mg is lost in leaching as a cation accompanying excess sulphate, the forest in the no-reduction scenario is gradually moving towards a situation where Mg becomes the growth-limiting element, Figure 11.10. Since the trees have a capacity for excess uptake of elements (Figure 8.1), this does not lead to an immediate leaching of nitrogen. In fact, with the parameterisation made in this study, this does not occur during the whole scenario. The improved scenario is, on the other hand, just enough to maintain the forest in a state of nitrogen-limitation for most of the time and, after a balancing act in a transition period, the forest settles down to a stable nitrogen-limited system. Interest in this kind of analysis is focussed along two lines of thought. Firstly, excess nitrogen will lead to leaching losses and the leached nitrogen may become a serious problem in lower-lying ground and surface waters. Secondly, there is concern that limitations on growth from other elements than nitrogen involve an increased vulnerability to stress factors such as frost and pathogens; forest trees have evolved under nitrogen limitation and have mechanisms to cope with stress factors associated with nitrogen limitations but only to a lesser extent with limitations by other elements.

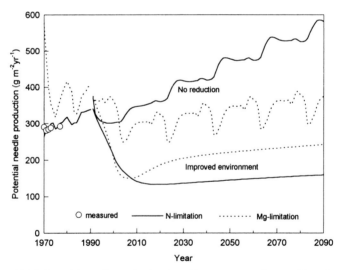

Figure 11.10 Potential needle production in a Norway spruce forest subjected to different deposition scenarios. During 1970-1990, depositions are actual depositions. From 1990, the depositions depict a no-reduction scenario and an improved environment scenario (from van Oene & Ågren 1995).

NOTES

[1] We follow a line of analysis suggested by McMurtrie et al. (1992), who, however, used canopy nitrogen concentration as independent variable.

[2] Data collected by Cole & Rapp (1981) show that the total amount of nitrogen can vary between at least 3000 and 40 000 kg ha^{-1} in forest ecosystems.

[3] Melin (1986).

[4] cf. Cole & Rapp (1981).

[5] Ågren (1989).

[6] Rolff & Ågren (199X), Ryan et al. (1996a).

[7] See Ryan et al. (1996ab) for further details on the model analysis; Axelsson & Bråkenhielm (1980) for description of the SWECON site and Beets & Brownlie (1987) for description of the BFG site.

[8] Bengtsson & Wikström (1993), Rolff & Ågren (199X).

[9] van Oene (1992), van Oene & Ågren (1995a).

12

Quality - the bridge between plant and soil

In this chapter we show how the abstract concept quality used in the previous chapters can be made an observable. By combining a range of decomposition studies of different litter types at different localities we demonstrate how quality can be estimated from conventional chemical fractionation. Finally, we show that the nitrogen productivity and quality are related. This completes the connection between plant properties and soil organic matter dynamics.

12.1 Components of quality

The idea behind the way we have defined quality is that it is a measure on the molecular scale of the accessibility to the carbon atom by the decomposer. There are a number of physico-chemical methods that can provide such information about a substrate. The most widely used technique is some form of chemical separation into fractions which are known to have different decomposabilities. This method has proved successful.[1]

Four components that can be obtained by a successive chemical fractionation of the litter are: water soluble, non-polar soluble, acid soluble, and acid insoluble (also often denoted lignin). The two first fractions are normally quite small so we combined them into an extractive fraction. These fractions represent substrates of different degradability and should, therefore, represent three different parts of the carbon spectrum, assuming that the technique used consistently extracts the same part of the quality spectrum. There is, of course, overlap in qualities whereby some range in quality may appear in two or more fractions. Improvements to fractionation techniques and decomposition data quality may, in the future, lead to a break-down of the spectrum into additional components.

If the relative amounts of the fractions in the litter are denoted c_{ex}, c_{as}, and c_{ai} ($c_{ex} + c_{as} + c_{ai} = 1$), and the carbon qualities associated with these fractions by q_{ex}, q_{as}, and q_{ai}, the average carbon quality is approximately given by

$$\bar{q}(t) = c_{ex}(t)q_{ex} + c_{as}(t)q_{as} + c_{ai}(t)q_{ai} \qquad (12.1)$$

We also need a scale for the qualities. We get this by arbitrarily setting $q_{ex} = 1$. If we knew the qualities of the other fractions, we could, by using the measured concentrations of the fractions (c_{ex},c_{as},c_{ai}), calculate from (4.50) how much of the litter should remain at any observation.

Berg (1981), Wessén & Berg (1986), Berg et al. (1987, 1991ab) and Aber et al. (1984) have gathered decomposition data for a range of litter types and site conditions. These data are, to our knowledge, the only more extensive data sets that follow changes in chemical fractions over an extended period of time. We have estimated q_{as} and q_{ai} as well as the other parameters by fitting measured amounts and the remaining carbon as given by (4.50) and with mean quality calculated from (4.52). With $q_{as} = 1.25$, $q_{ai} = 0.65$, $e_0 = 0.25$, and $\eta_{11} = 0.36$ we obtained the correlation shown in Figure 12.1.

From (4.52) we see that a plot of $\left(q_0/\bar{q}\right)^{\beta} - 1$ versus time gives a straight line with slope $\beta\eta_{11}fc_{u0}q_0^{\beta}$. One such plot is shown in Figure 12.2 for *Pinus sylvestris* needles. High values of β (7–10) are necessary to transform the data into linear form. We have chosen $\beta = 7$ for the further analysis. Larger values give an infinite accumulation of soil carbon with a constant litter input rate (4.53). We have repeated this analysis for other data sets with *Pinus sylvestris*, but where it is decomposing at other sites. Since the litter qualities are similar, q_0 is constant, and the decomposer communities presumably not too different, the main factors that should cause a difference in the slopes are climatic. In Figure 12.3 we show a regression of the specific decomposition rate k with the different u_0s estimated as in Figure 12.2 versus mean annual air temperature for six sites where this information is available. There is very good correlation between temperature and k ($r^2 = 0.82$). The increase in rate with temperature is steep, going from 1 °C to 9 °C increases the rate by almost a factor of 3, which is consistent with other studies of soil respiration.[2]

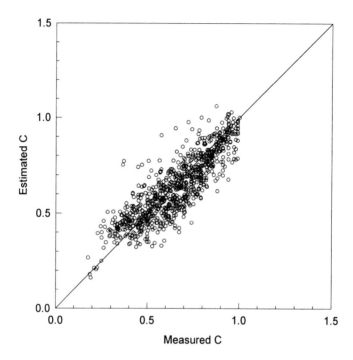

Figure 12.1 Estimated remaining carbon and measured remaining carbon. The data sets represent 19 litter types and 16 localities. 978 data points (from Ågren & Bosatta 1996).

We are not aware of any direct measurements of $u_0q_0^\beta$ but Anderson & Domsch (1990) observed qCO_2 (microbial respiration per unit biomass and time) in the range $1.7 \cdot 10^{-4}\ h^{-1} - 17.0 \cdot 10^{-4}\ h^{-1}$ for a range of agricultural soils and with microbial biomasses around 2.5% of soil carbon. With an assumed efficiency of 0.25 this corresponds to a microbial growth rate per unit of carbon of $4 \cdot 10^{-5} - 4 \cdot 10^{-4}\ d^{-1}$ which can be compared with the observed range of values for $u_0q_0^\beta$ $4.4 \cdot 10^{-4} - 12 \cdot 10^{-4}\ d^{-1}$ (0.16–0.43 yr^{-1}).

The parameters we estimate are strongly correlated and several other parameter combinations give an almost identical fit. When, for example, we tried to add a term linear in q to $e(q)$, (4.41), there was almost no difference to the fit, and the parameter values were such that we found only a small variation in $e(q)$ over the quality interval of interest.[3]

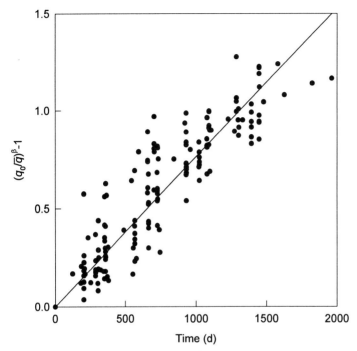

Figure 12.2 Evolution of $(q_0/\bar{q})^\beta - 1$ over time for *Pinus sylvestris* needles (from Ågren & Bosatta 1996).

It should also be observed that with the present analysis we cannot separate e_0 and η_{11} from each other. It is, however, clear that the three quality parameters have to rank as $q_{ai} < q_{ex} < q_{as}$. That $q_{ex} < q_{as}$ is contrary to assumptions used in some ecosystem models[4] and may be caused by certain highly resistant compounds in the non-polar fraction. It is also consistent with the observation that some plant material originating from the extractive fraction can be transported long distances as DOC before being degraded.

When we compare ks estimated directly from (4.10) with measured ones in the Black Hawk Island study[5] there is no overall agreement because (4.10) overestimates the decomposition rates of the bark, root, and woody litters. However, these litters have a different physical shape with a surface-to-volume ratio that is approximately a factor of 10 less than that of the leaf litters. It is, therefore, possible that we overestimate the initial accessibility of the carbon to the decomposers in these litters. There are also tendencies for the specific decomposition rate to increase with time in some of these litters. By assuming that there is indeed a physical difference between the litters, and by decreasing u_0 by a factor of 4 for these three litter types, we get a good correspondence between calculated and measured ks. A regression through the origin has an $r^2 = 0.76$.

Nowhere in the calculation of decomposition rates have we made any explicit references to the nitrogen content of the material, in contrast to other studies that emphasise nitrogen as a key indicator of decomposability. However, with our partitioning of the substrate, differences in nitrogen concentrations between fractions can account for this. Studies that carefully follow the development of nitrogen in the different chemical fractions are required to illuminate this question. It should also be observed that differences in nitrogen concentrations are sometimes observed to have smaller effects on decomposition rates within a litter type than the same differences between litter types.[6]

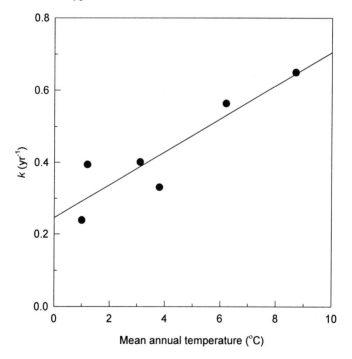

Figure 12.3 Specific decomposition rate for *Pinus sylvestris* needles versus mean annual air temperature (from Ågren & Bosatta 1996).

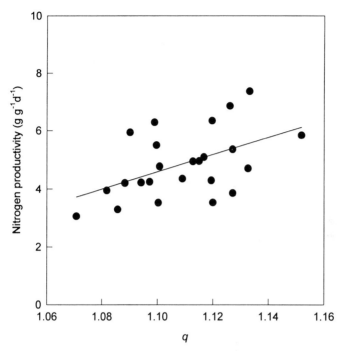

Figure 12.4 Relation between nitrogen productivity and quality. The regression line has $r^2 = 0.25$. Based on data from Poorter & Bergkotte (1992).

12.2   Plants and quality

While it is easy to understand the concept of quality with reference to decomposers it is less obvious how to combine it with the plant properties suggested earlier in this book. However, as discussed extensively by Penning de Vries (1974) the various compounds forming the plant biomass represent different energetic costs. This should also be reflected in plant performance. Most studies of this kind have looked at the question of plant chemical composition as a question of defence against herbivory and pathogens. Niemann et al. (1992) showed that R_W correlated with pyrolysis mass spectrometry information, and Poorter & Bergkotte (1992) found a correlation between R_W and the chemical fractions used by us to define litter quality. From the later study, we can therefore calculate $\bar{q}$ with (12.1); we have used ash-free concentrations. Since they also give the plant nitrogen concentrations, we can also estimate P_N under the assumption that $c_{N,min} = 0$ and that excess uptake is negligible. Figure 12.4 shows the correlation between P_N and $\bar{q}$. There is, thus, a relation between quality as percieved by decomposers and plant properties whereby substrates that permit high decomposer growth rates (high q) are produced by plants that have potentially high growth rates (high P_N).

NOTES

[1] Ågren & Bosatta (1996). Alternative methods are NMR- (Baldock et al. 1992, Beyer et al. 1993, Hempfling et al. 1987, Nordén & Berg 1990) and IR- (Gillon et al. 1993) spectroscopy. We have also tried to analyse detailed NMR data of decomposing litters, but without success. The problem in using spectroscopic methods, we think, is that there are no simple relationships between measured quantities and qualities perceived by decomposers, or that the techniques are still too crude. Spectroscopic methods give a very fine-scale description of the molecules, but to the decomposer it is also important what the rest of that molecule looks like when attacking, e.g. a particular methyl group.

[2] E.g. Peterjohn et al. (1994).

[3] This may seem at odds with observations of variations in decomposer efficiency (e.g. Hart et al. 1994) but might be realistic as a long-term average, cf. the last three-quarters of Hart et al.'s observations.

[4] Rastetter et al. (1991).

[5] Aber et al. (1984).

[6] Berg & Ekbohm (1991).

Epilogue

Although we are now at the end of the book, we are only at the beginning of theoretical ecosystem ecology. We believe that we have been able to demonstrate that a single equation (1.1) can form the basis for a fruitful analysis of this scientific domain. Equation (1.1) is a linear equation. An important difference between linear and non-linear systems is that superposition can be applied to the former but not to the latter. In linear systems, the effect of the combined effects of two different causes is merely the superposition of the effects of each cause taken individually. In non-linear systems, adding a small cause to one that is already present can induce dramatic effects that have no common measure with the amplitude of the cause. Equation (1.1) can be seen as a first approximation (the linear terms) to a more general equation including non-linearities. The interesting and critical question is under which circumstances the property of superposition breaks down, i.e. under which conditions do the non-linearities have dramatic effects upon the properties of the ecosystem?

One example on non-linearities was briefly mentioned in Chapter 6, namely the possibility of chemical reactions between secondary products of decomposition. Suppose that a bimolecular reaction between components q' and q'' of the spectrum is occurring at a rate $K(q',q'')$. We then add the following non-linear term to (4.7)

$$\int_0^\infty \int_0^\infty [D(q,q',q'') - \delta(q - q')] K(q',q'') \rho_C(q',t) \rho_C(q'',t) dq' dq''$$

where $D(q,q',q'')$ denotes the fraction of reaction products that are of quality q.

This example can be extended to reactions of any degree and shows how our formalism (and in particular the dispersion function) can be generalised to include non-linearities. Other non-linearities that we have introduced, e.g. (9.5), are simpler to analyse and have simpler consequences.

Equation (1.1) can also be generalised to include spatial co-ordinates such as soil depth. In this case, the density functions also become functions of spatial co-ordinates, $\rho(q,z,t)$, and the system of ordinary differential equations for the moments of the density distribution (4.26-4.28 or A2.5) becomes a system of partial differential equations in time and depth. This generalisation has been shown useful for analysing the problems of distribution and mineralisation of C, N, P, and S in soil profiles (Bosatta & Ågren 1996). We have also shown how the formalism can describe isotope discrimination during decomposition (Ågren et al., 1996).

Many problems remain to be solved. The models for decomposer properties need to be further developed and critical experiments in this field are necessary.

A more extensive analysis of relationships between quality and plant performance, not only of potential plant performance but also of realised performance such as variations in plant nitrogen concentration, is essential.

The major endeavours must, however, be made at the ecosystem level where we feel that only the surface has been scratched so far. Surprises in this field are almost inevitable. It might even become necessary to introduce new concepts and not only try to construct ecosystems out of the underlying ones.

Appendices

A.1 Some properties of the delta function

One of the possible definitions of a delta function is

$$\delta(q - q_0) = \lim_{\sigma \to 0} \frac{1}{\sigma\sqrt{\pi}} e^{-\frac{(q-q_0)^2}{\sigma^2}} \tag{A1.1}$$

The characteristics of the delta function is that it is zero everywhere, except at the point q_0, where it goes to infinity.

Some useful mathematical properties are

$$\int_{q_0-\varepsilon}^{q_0+\varepsilon} \delta(q - q_0) dq = 1, \quad \varepsilon > 0 \tag{A1.2}$$

$$\delta(-q) = \delta(q) \tag{A1.3}$$

$$\frac{d\delta(q)}{dq} = -\frac{d\delta(-q)}{dq} \tag{A1.4}$$

$$\int_{q_0-\varepsilon}^{q_0+\varepsilon} f(q)\delta^{(n)}(q - q_0) dq = (-1)^n f^{(n)}(q_0) \tag{A1.5}$$

where (n) denotes the nth derivative.

Let $h(q)$ and $u(q)$ be arbitrary functions. Using partial integration:

$$\int_{q_0-\varepsilon}^{q_0+\varepsilon} h(q)\frac{\partial}{\partial q}\left[u(q)\delta(q - q_0)\right] dq = \int_{q_0-\varepsilon}^{q_0+\varepsilon} \frac{dh(q)}{dq} u(q_0)\delta(q - q_0) dq = \frac{dh(q_0)}{dq} u(q_0)$$

$$\tag{A1.6}$$

$$\delta[g(q)] = \sum_n \frac{1}{|g'(q_n)|} \delta(q - q_n), \quad \text{where } g(q_n) = 0 \text{ and } g'(q_n) \neq 0$$

$$(A1.7)$$

A.2   Numerical solutions to (4.7) and to the moment expansion

Given $e(q)$, $u(q)$, and $D(q,q')$ the numerical integration of equation (4.7) is not difficult, although it requires considerable computer resources due to the small spacing in q and t necessary for satisfactory accuracy. Figure A2.1 shows the distribution $\rho_C(q,t)$ generated by a numerical solution to (4.7) with an initially log-normally, but very narrowly, distributed ρ_C and D likewise given by a log-normal distribution

$$D(q,q') = \frac{1}{q(2s^2\pi)^{1/2}} e^{-\frac{[\ln q - \ln(q'-\varepsilon)]^2}{2s^2}} \tag{A2.1}$$

$$e(q) = e_1 q \tag{A2.2}$$

$$u(q) = u_0 q^\beta \tag{A2.3}$$

As time evolves the main features are:

i) the original distribution $\rho_C(q,0)$ decreases in height due to mineralisation of carbon and new peaks separated by distances ε begin to appear. Such peaks will be visible only if ε is large relative to s, i.e. for each cycle of carbon through the microbial biomass the quality is on the average displaced a large distance relative to width of the distribution D. In the limit of $s \to 0$ the carbon distribution $\rho_C(q,t)$ will consist of a series of δ-functions;

ii) the centre of the distribution, $\hat{q}(t)$, shifts to lower values; and

iii) the breadth of the distribution, $\sigma^2(t)$, increases at first but passes over a maximum after which it also decreases in time.

Including more of the linear terms in the moment expansion of the equations (4.26)-(4.28) we get:

$$\frac{dC(t)}{dt} = -\int k(q)\rho_C(q,t)dq = -\left[k(\hat{q}) + \frac{\sigma^2(t)}{2} \frac{\partial^2}{\partial \hat{q}^2} k(\hat{q}) + \frac{\mu_3}{6} \frac{\partial^3}{\partial \hat{q}^3} k(\hat{q}) + \cdots \right] C(t)$$

$$(A2.4)$$

$$\frac{d\hat{q}(t)}{dt} = -\frac{dC(t)}{dt}\frac{\hat{q}(t)}{C(t)} - \frac{1}{C(t)}\int q[k(q) + f_C u(q)]\rho_C(q,t)dq +$$

$$+ \frac{f_C}{C(t)}\int[q - \eta_1(q)]u(q)\rho_C(q,t)dq =$$

$$= -f_C u(\hat{q})\eta_1(\hat{q}) - \left[\frac{\partial}{\partial\hat{q}}k(\hat{q}) + \frac{1}{2}f_C\frac{\partial^2}{\partial\hat{q}^2}(u(\hat{q})\eta_1(\hat{q}))\right]\sigma^2(t) -$$

$$- \left[\frac{1}{2}\frac{\partial^2}{\partial\hat{q}^2}k(\hat{q}) + \frac{1}{6}f_C\frac{\partial^3}{\partial\hat{q}^2}(u(\hat{q})\eta_1(\hat{q}))\right]\mu_3(t) + \dots \qquad (A2.5)$$

$$\frac{d\sigma^2(t)}{dt} = -\frac{dC(t)}{dt}\frac{\sigma^2(t)}{C(t)} - \frac{1}{C(t)}\int k(q)[q - \hat{q}(t)]\rho_C(q,t)dq +$$

$$+ \frac{1}{C(t)}f_C\int\eta_2(q)u(q)\rho_C(q,t)dq - \frac{2f_C}{C(t)}\int\eta_1(q)[q - \hat{q}(t)]u(q)\rho_C(q,t)dq =$$

$$= f_C u(\hat{q})\eta_2(\hat{q}) - \left[2f_C\frac{\partial}{\partial\hat{q}}(u(\hat{q})\eta_1(\hat{q})) - \frac{1}{2}f_C\frac{\partial^2}{\partial\hat{q}^2}(u(\hat{q})\eta_2(\hat{q}))\right]\sigma^2(t) -$$

$$- \left[\frac{\partial}{\partial\hat{q}}k(\hat{q}) - \frac{1}{6}f_C\frac{\partial^3}{\partial\hat{q}^3}(u(\hat{q})\eta_2(\hat{q})) + \frac{\partial^2}{\partial\hat{q}^2}(u(\hat{q})\eta_1(\hat{q}))\right]\mu_3(t) + \dots$$

$$(A2.6)$$

$$\frac{d\mu_3(t)}{dt} = -f_C u(\hat{q})\eta_3(\hat{q}) +$$

$$+ f_C\left[-\frac{1}{2}\frac{\partial^2}{\partial\hat{q}^2}(u(\hat{q})\eta_3(\hat{q})) + 3\frac{\partial}{\partial\hat{q}}(u(\hat{q})\eta_2(\hat{q})) - 3u(\hat{q})\eta_1(\hat{q})\right]\sigma^2(t)$$

$$+ f_C\left[-\frac{1}{6}\frac{\partial^3}{\partial\hat{q}^3}(u(\hat{q})\eta_3(\hat{q})) + \frac{3}{2}\frac{\partial^2}{\partial\hat{q}^2}(u(\hat{q})\eta_2(\hat{q})) - 3\frac{\partial}{\partial\hat{q}}(u(\hat{q})\eta_1(\hat{q}))\right]\mu_3(t) + \dots$$

$$(A2.7)$$

where all integrals extend over the whole spectrum of qualities, 0 to ∞.

In Figure A2.2 the values of $C(t)$, $\hat{q}(t)$, $\sigma^2(t)$, and $\mu_3(t)$ calculated from the numerical solution (Figure A2.1) are compared with the corresponding values generated with the use of equations (A2.4)-(A2.6). With the parameter values chosen and over the time interval studied, the total carbon is virtually unaffected by the number of terms included in the expansion. On the other hand, the moments are clearly affected and increasing the number of terms improves the agreement between the expansion and the "exact" solution. As can be seen for the variance, there seems to come a point where the inclusion of three moments does worse than just two moments. However, if the calculations had been extended over a longer period this impression would probably change. Experimentally, it is generally only the total carbon that is studied. Its weak dependence on the level of approximation indicates that it is an insensitive variable with respect to the underlying processes. Therefore, if we want to study decomposition processes at greater depth we need to investigate other variables.

The shape of the carbon distribution is modified by two forces. One is the differential respiration rate of carbon of different qualities, which appears in terms of the form $\mu_n \partial^m k / \partial q^m$. The other is the redistribution of carbon over the quality spectrum due to microbial assimilation, which appears in terms of the form $\partial^m (u\eta_n)/\partial q^m$. For the best convergence of the moment expansion these two types of terms should be of equal magnitude.

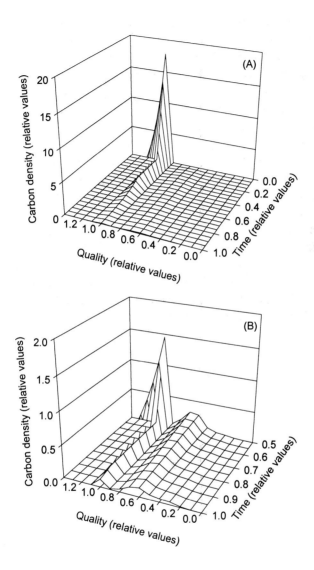

Figure A2.1 An example of a numerical solution of the temporal development of the distribution of carbon over substrate quality, equation (4.7) with the models defined by (A2.1)-(A2.3), $e = 0.3$, $2s^2 = 0.05$, $\beta = 3$, $e_1 = 0.2$. (A) The entire simulation period $0 < t < 1$. (B) The last part of the simulation period $0.5 < t < 1$ (from Bosatta & Ågren 1991a).

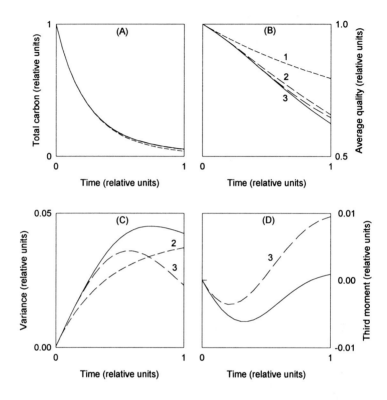

Figure A2.2 Comparison of results from the numerical solution of equation (4.7) and the corresponding variables calculated with the moment expansion, equations (A2.4)-(A2.6). Solid line = numerical solution. Broken lines = moment expansion, where the number on the curve indicates how many moments have been included (1 = only $\hat{q}(t)$, 2 = $\hat{q}(t)$ and s^2; 3 = $\hat{q}(t)$, s^2 and μ^3). Parameters as in Figure A2.1. (A) Total carbon. The three curves are indistinguishable from each other. (B) Average carbon quality. (C) Variance of carbon quality. (D) Third moment of carbon distribution (from Bosatta & Ågren 1991a).

References

Aber, J.D. & Melillo, J.M. 1980. Litter decomposition: measuring relative contributions of organic matter and nitrogen to forest soils. *Canadian Journal of Botany* 58:416-421.

Aber, J.D. & Melillo, J.M. 1982a. *FORTNITE: A computer model of organic matter and nitrogen dynamics in forest ecosystems*. College of Agricultural and Life Sciences, University of Wisconsin-Madison, Research Report R3130, 49 pp.

Aber, J.D. & Melillo, J.M. 1982b. Nitrogen immobilization in decaying leaf litter as a function of initial nitrogen and lignin content. *Canadian Journal of Botany* 60:2263-2269.

Aber, J.D., McClaugherty, C.A. & Melillo, J.M. 1984. *Litter decomposition in Wisconsin forests - Mass loss, organic-chemical constituents and nitrogen*. College of Agricultural and Life Sciences, University of Wisconsin-Madison, Research Report R3284, 72 pp.

Aber, J.D., Melillo, J.M., Nadelhofer, K.J., McClaugherty, C.A. & Pastor, J. 1985. Fine root turnover in forest ecosystems in relation to quantity and form of nitrogen availability: a comparison of two methods. *Oecologia (Berlin)* 66:317-321.

Aber, J.D., Magill, A., Boone, R., Melillo, J.M., Steudler, P. & Bowden, R. 1993. Plant and soil response to chronic nitrogen aditions at the Harvard forest, Massachusetts. *Ecological Applications* 3:156-166.

Abramowitz, M. & Stegun, I.A. 1972. *Handbook of Mathematical Functions*. Dover, New York.

Ågren, G.I. 1983. Nitrogen productivity of some conifers. *Canadian Journal of Forest Research* 13:494-500.

Ågren, G.I. 1985a. Theory for growth of plants derived from the nitrogen productivity concept. *Physiologia Plantarum* 64:17-28.

Ågren, G.I. 1985b. Limits to plant production. *Journal of Theoretical Biology* 113:89-92.

Ågren, G.I. 1988. Ideal nutrient productivities and nutrient proportions in plant growth. *Plant, Cell and Environment* 11:613-620.

Ågren, G.I. 1989. Willow stand growth and seasonal nitrogen turnover. In: Perttu, K.L. & Kowalik, P.J.(eds.) *Modelling of Energy Forestry Growth, Water Relations and Economics.* Simulation Monographs, Pudoc, Wageningen. pp. 77-85.

Ågren, G.I. 1996. Nitrogen productivity of photosynthesis minus respiration to calculate plant growth. *Oikos* 76:529-535.

Ågren, G.I. & Axelsson, B. 1980. Population respiration: A theoretical approach. *Ecological Modelling* 11:39-54.

Ågren, G.I. & Bosatta, E. 1987. Theoretical analysis of the long-term dynamics of carbon and nitrogen in soils. *Ecology* 68:1181-1189.

Ågren, G.I. & Bosatta, E. 1988. Nitrogen saturation of terrestrial ecosystems. *Environmental Pollution* 54:185-197.

Ågren, G.I. & Bosatta, E. 1990. Theory and model or art and technology in ecology. *Ecological Modelling* 50:213-220.

Ågren, G.I. & Bosatta, E. 1996. Quality: A bridge between theory and experiment in soil organic matter studies. *Oikos* 76:522-528.

Ågren, G.I. & Ingestad, T. 1987. Root:shoot ratio as a balance between nitrogen productivity and photosynthesis. *Plant, Cell and Environment* 10:579-586.

Ågren, G.I., Andersson, F. & Fagerström, T. 1980. Experiences of ecosystem research in the Swedish Coniferous Forest Project. In: Persson, T. (ed.) *Structure and Function of Northern Coniferous Forests - An Ecosystem Study, Ecological Bulletins* 32:591-596.

Ågren, G.I., McMurtrie, R., Pastor, W.J., Pastor, J. & Shugart, H.H. 1991. State-of-the-art of models of production-decomposition linkages in conifer and grassland ecosystems. *Ecological Applications* 1:118-138.

Ågren, G.I., Johnson, D.W., Kirschbaum, M. & Bosatta, E. 1996. Ecosystem physiology- Soil organic matter. In: Breymeyer, A., Hall, D.O., Melillo, J.M. & Ågren, G.I. (eds.) *Global Change: Effects on Forests and Grasslands*, J.Wiley, Chichester pp. 207-228.

Ågren, G.I., Bosatta, E. & Balesdent, J. 1996. Isotope discrimination during decomposition of organic matter - a theoretical analysis. *Soil Science Society of America Journal* 60:1121-1126.

Ajtay, G.L., Ketner, P. & Duvigneaud, P. 1979. Terrestrial primary production and phytomass. In: Bolin, B., Degens, E.T., Kempe, S. & Ketner, P. (eds.) 1979. *The Global Carbon Cycle* (SCOPE report; 13). John Wiley & Sons, Chichester. pp. 129-181.

Allen, T.F.H. & Starr, T.B. 1982. *Hierarchy; Perspectives for Ecological Complexity.* The University of Chicago Press, Chicago.

Allison 1973. *Soil Organic Matter and its Role in Crop Production.* Elsevier, Amsterdam.

Anderson, T.-H. & Domsch, K.H. 1989. Ratios of microbial biomass carbon to total organic carbon in arable soils. *Soil Biology and Biochemistry* 21:471-479.

Anderson, T.-H. & Domsch, K.H. 1990. Application of eco-physiological quotients (qCO2 and qD) on microbial biomasses from soils of different cropping histories. *Soil Biology and Biochemistry* 22:251-255.

Andersson, F., Axelsson, B., Lohm, U., Perttu, K. & Ågren, G.I. 1980. Skogen som miljö - grundforskning om landekosystem. *NFR:s årsbok 1979/80.* pp. 141-179.

Axelsson, G. & Berg, B. 1988. Fixation of ammonia (^{15}N) to Pinus silvestris needle litter in different stages of decomposition. *Scandinavian Journal of Forest Research* 3:273-279.

Axelsson, B. & Bråkenhielm, S. 1980. Investigation sites of the Swedish Coniferous Forest Project - biological and physiogeographical features. In: Persson, T. (ed.) *Structure and Function of Northern Coniferous Forests - An Ecosystem Study, Ecological Bulletins* (Stockholm) 32:25-64.

Baldock, J.A., Oades, J.M., Waters, A.G., Peng, X., Vassallo, A.M. & Wilson, M.A. 1992. Aspects of the chemical structure of soil organic materials as revealed by solid-state 13C NMR spectroscopy. *Biogeochemistry* 16:1-42.

Barber, B.L. & Van Lear, D.H. 1984. Weight loss and nutrient dynamics in decomposing woody loblolly pine logging slash. *Soil Science Society of America Journal* 48:906-910.

Bartholomew, W.V. 1965. Mineralization and immobilization of nitrogen in the decomposition of plant and animal residues. In: Bartholomew, W.V. & Clark, F.E. (eds.) *Soil Nitrogen. Agronomy Monographs* 10. ASA and SSSA, Madison, WI. pp. 285-306.

Baule, B. 1956. Das Ertragsgesetz als Mittel zur Bestimmingung der Bodennährstoffe. *Zeitschrift für Acker- und Pflanzenbau* 100:261-272.

Beets, P.N. & Brownlie, R.K. 1987. Puruki experimental catchment: site, climate, forest management, and research. *New Zealand Journal of Forest Science* 17:137-160.

Beyer, L., Schulten, H R., Freund, R. & Irmler, U. 1993. Formation and properties of organic matter in a forest soil, as revealed by its biological activity, wet chemical analysis, CPMAS ^{13}NMR. *Soil Biology and Biochemistry* 25: 587-596.

Bengtsson, J. & Wikström, F. 1993. Effects of whole-tree harvesting on the amount of soil carbon: Model results. *New Zealand Journal of Forestry Science* 23:380-389.

Berendse, F & Aerts, R. 1987. Nitrogen-use-efficiency: A biologically meaningful definition? *Functional Ecology* 1:293-296.

Berg, B. 1981. Litter decomposition studies within SWECON. Data on weight loss and organic chemical composition. *Swedish Coniferous Forest Project Internal Report* 101.

Berg, B. 1984. Decomposition of root litter and some factors regulating the process: Long-term root litter decomposition in a Scots pine forest. *Soil Biology and Biochemistry* 16:609-617.

Berg, B. 1988. Dynamics of nitrogen (^{15}N) in decomposing Scots pine (Pinus sylvestris) needle litter. Long-term decomposition in a Scots pine forest. VI. *Canadian Journal of Botany* 66:1539-1546.

Berg, B. & Ågren, G.I. 1984. Decomposition of needle litter and its organic chemical components. - Theory and field experiments. Long-term decomposition in a Scots pine forest. IV. *Canadian Journal of Botany* 62:2880-2888.

Berg, B. & Ekbohm, G. 1983. Nitrogen immobilization in decomposing needle litter at variable carbon:nitrogen ratios. *Ecology* 64:63-67.

Berg, B. & Ekbohm, G. 1991. Litter mass-loss rates and decomposition patterns in some needle and leaf litter types. Long-term decomposition in a Scots pine forest. VII. *Canadian Journal of Botany* 69:1449-1456.

Berg, B. & McClaugherty, C.A. 1989. Nitrogen and phosphorus release from decomposing litter in relation to the disappearance of lignin. *Canadian Journal of Botany* 67:1148-1156.

Berg, B. & Staaf, H. 1980. Decomposition rate and chemical changes in decomposing needle litter of Scots pine. II. Influence of chemical composition. In: Persson, T. (ed.) *Structure and Function of Northern Coniferous Forests - An Ecosystem Study, Ecological Bulletins* (Stockholm) 32:373-390.

Berg, B. & Staaf, H. 1981. Leaching, accumulation and release of nitrogen in decomposing forest litter. In: Clark, F.E. & Rosswall, T. (eds.) *Terrestrial Nitrogen Cycles, Ecological Bulletins* (Stockholm) 33:163-178.

Berg, B., Jansson, P.-E. & Meentemayer, V. 1984. Litter decomposition and climate - Regional and local models. In: Ågren, G.I. (ed.) State and Change of Forest Ecosystems - Indicators in Current Research. *Swedish University of Agricultural Sciences, Department of Ecology & Environmental Research, Report* nr 13, pp. 389-404.

Berg, B., Müller, M. & Wessén, B. 1987. Decomposition of red clover (*Trifolium pratense*) roots. *Soil Biology and Biochemistry* 19:589-593.

Berg, B., Booltink, H., Breymeyer, A., Ewertsson, A., Gallardo, A., Holm, B., Johansson, M.B., Kuovioja, S., Meentemeyer, V., Nyman, P., Olofsson, J., Reurslag, A., Staaf, H., Staaf, I. & Uba, L. 1991a. Data on needle litter decomposition and soil climate as well as site characteristics for some coniferous forest sites. Part 1. Site characteristics. *Department of Ecology and Environmental Research, Swedish University of Agricultural Sciences Report* 41.

Berg, B., Booltink, H., Breymeyer, A., Ewertsson, A., Gallardo, A., Holm, B., Johansson, M.B., Kuovioja, S., Meentemeyer, V., Nyman, P., Olofsson, J., Reurslag, A., Staaf, H., Staaf, I. & Uba, L. 1991b. Data on needle litter decomposition and soil climate as well as site characteristics for some coniferous forest sites. Part 2. Decomposition data. *Department of Ecology and Environmental Research, Swedish University of Agricultural Sciences Report* 42.

Bergner, P.E. 199X. Reflections on theory and linguistics. (Manuscript submitted to *Journal of Biological Systems*).

Bertrand, D. 1950. Survey of contemporary knowledge of biogeochemistry. 2. The biogeochemistry of vanadium. *Bulletin of the American Museum of Natural History* 94:403-456.

Biederbeck, V.O. 1978. Soil organic sulfur and fertility. In: Schnitzer, M. & Khan, S.U. (eds.) *Soil Organic Matter. Developments in Soil Science 8,* Elsevier, Amsterdam. pp. 273-310.

Bockheim J.G, Leide J.E & Tavella D.S. 1986. Distribution and cycling of macronutrients in a Pinus resinosa plantation fertilized with nitrogen and potassium. *Canadian Journal of Forest Research* 16:778-785.

Bolin, B. 1979. On the role of the atmosphere in biogeochemical cycles. *Quarterly Journal of the Royal Meteorological Society* 105:25-42.

Bosatta, E. & Ågren, G.I. 1985. Theoretical analysis of decomposition of heterogeneous substrates. *Soil Biology and Biochemistry* 17:601-610.

Bosatta & Ågren, G.I. 1991a. Dynamics of carbon and nitrogen in the organic matter of the soil: A generic theory. *The American Naturalist* 138:227-245.

Bosatta, E. & Ågren, G.I. 1991b. Theoretical analysis of carbon and nutrient interactions in soils under energy-limited conditions. *Soil Science Society of America Journal* 55:728-733.

Bosatta, E. & Ågren, G.I. 1994. Theoretical analysis of microbial biomass dynamics in soils. *Soil Biology and Biochemistry* 26:143-148.

Bosatta, E. & Ågren, G.I. 1995a. Theoretical analyses of the interactions between inorganic nitrogen and soil organic matter. *European Journal of Soil Science* 46:109-114.

Bosatta, E. & Ågren, G.I. 1995b. The power and reactive continuum models as particular cases of the q-theory of organic matter dynamics. *Geochimica et Cosmochimica Acta* 59:3833-3835.

Bosatta, E. & Ågren, G.I. 1996. Theoretical analyses of carbon and nutrient distribution and dynamics in soil profiles. *Soil Biology and Biochemistry* 28:1523-1531.

Bosatta, E. & Berendse, F. 1984. Energy or nutrient regulations of decomposition: Implications for the mineralization-immobilization response to perturbations. *Soil Biology and Biochemistry* 16:63-67.

Bosatta, E. & Staaf, H. 1982. The control of nitrogen turn-over in forest litter. *Oikos* 39:143-151.

Bottner, P, Mneimne, Z. & Billes, G. 1984. Réponse de la biomass microbienne a l'adjonction au sol de matériel végétal marque au ^{14}C: Rôle des racines vivantes. *Soil Biology and Biochemistry* 16:305-314.

Bunge, M. 1967. *Scientific Research I. The Search for System.* Studies in the Foundations. Methodology and Philosophy of Science, Vol.3/i. Springer-Verlag, Berlin.

Burge, W.D. & Broadbent, F.E. 1961. Fixation of ammonia by organic soils. *Soil Science Society of America Proceedings* 25:199-204.

Burns I.G. 1994. Studies of the relationship between the growth rate of young plants and their total-N concentration using nutrient interruption techniques: Theory and experiment. *Annals of Botany* 74:143-157.

Byrd, G. T., Sage, R. F. & Brown, R. H. 1992. A comparison of dark respiration between C3 and C4 plants. *Plant Physiology* 100:191-198.

Carter, J.N., Bennet, O.L. & Pearson, R.W. 1967. Recovery of fertilizer nitrogen under field conditions using Nitrogen-15. *Soil Science Society of America Proceedings* 31:50-56.

Cerri, C. & Jenkinson, D.S. 1981. Formation of microbial biomass during the decomposition of ^{14}C labelled ryegrass in soil. *Journal of Soil Science* 32:619-626.

Chapin, F.S. 1980. The mineral nutrition of wild plants. *Annual Review of Ecology and Systematics* 11:233-260.

Charles-Edwards, D.A. 1981. *The Mathematics of Photosynthesis and Productivity.* Academic Press, London.

Cheshire, M.V., Inkson, R.H.E., Mundie, C.M. & Sparling, G.P. 1988. Studies on the rate of decomposition of plant residues in soil following changes in sugar components. *Journal of Soil Science* 39:227-236.

Cole, D.W. & Rapp, M. 1981. Elemental cycling in forest ecosystems. In: Reichle, D.E. (ed.) *Dynamic Properties of Forest Ecosystems* (International Biological Programme; 23). Cambridge University Press, Cambridge. pp. 341-409.

Cole, C.V., Innis, G.S. & Stewart, J.W.B. 1977. Simulation of phosphorus cycling in semiarid grasslands. *Ecology* 58:1-15.

Comins H.N. & McMurtrie R.E. 1993. Long-term response of nutrient-limited forest to CO_2 enrichment; Equilibrium behavior of plant-soil models. *Ecological Applications* 3:666-681.

Constanza, R., Wainger. L., Folke, C. & Mäler, K.-G. 1993. Modeling complex ecological economic systems. *BioScience* 43:545-555.

Dalal, R.C. 1977. Soil organic phosphorus. *Advances in Agronomy* 29:83-117.

Duncan, W.G., Loomis, R.S., Williams, W.A. & Hanua, R. 1967. A model for simulating photosynthesis in plant communities. *Hilgardia* 38:181-205.

Ellert, B.H. & Bettany, J.R. 1988. Comparison of kinetic models for describing net sulfur and nitrogen mineralization. *Soil Science Society of America Journal* 52:1692-1702.

Epstein, E. 1972. *Mineral Nutrition of Plants: Principles and Perspectives.* John Wiley & Sons, New York.

Ericsson, T. 1995. Growth and shoot:root ratio in seedlings in relation to nutrient availability. *Plant and Soil* 168/169:205-214.

Ericsson, T. & Ingestad, T. 1988. Nutrition and growth of birch seedlings at varied relative phosphorus addition rates. *Physiologia Plantarum* 72:227-235.

Ericsson, T., Larsson, C.-M. & Tillberg, E. 1982. Growth response of Lemna to different levels of nitrogen nutrition. *Zeitschrift für Pflanzenphysiologie* 105:331-340.

Farquhar, G. D. & von Caemmerer, S. 1982. Modelling of photosynthetic response to environmental conditions. In: Lange, O.L., Nobel, P.S., Osmond, C.B. & Ziegler, H. (eds.) *Physiological Plant Ecology. II. Water Relations and Carbon Assimilation. Encyclopaedia of Plant Physiology, New Series* 12B. Springer-Verlag, Berlin. pp. 549-587.

Field, C. 1983. Allocating leaf nitrogen for the maximization of carbon gain: Leaf age as a control on the allocation program. *Oecologia(Berlin)* 56:341-347.

Field, C. & Mooney, H.A. 1986. The photosynthesis-nitrogen relationship in wild plants. In: Givnish, T. (ed.) *On the Economy of Plant Form and Function.* Cambridge University Press, Cambridge. pp. 23-55.

Field C.B., Chapin III, F.S., Matson, P.A. & Mooney, H.A. 1992. Responses of terrestrial ecosystems to the changing atmosphere: a resource-based approach. *Annual Review of Ecology and Systematics* 23:201-235.

Fog, K. 1988. The effect of added nitrogen on the rates of decomposition of organic matter. *Biological Review* 63:433-462.

Fogel, R. & Cromack, K. Jr. 1977. Effect of habitat and substrate quality on Douglas fir litter decomposition in Western Oregon. *Canadian Journal of Botany* 55:1632-1640.

Gauch, H.G. 1972. *Inorganic Plant Nutrition.* Dowden, Hutchinson & Ross, Inc. Stroudsburg, Pa.

Gillon, D., Joffre, R. & Dardenne, P. 1993. Predicting the stage of decay of decomposing leaves by near infrared reflectance spectroscopy. *Canadian Journal of Forest Research* 23:1552-2559.

Glansdorff, P. & Prigogine, I. 1971. *Structure, Stabilité et Fluctuation.* Masson et Cie. Paris.

Gorham, E., Vitousek, P.M. & Reiners, W.A. 1979. The regulation of chemical budgets over the course of terrestrial ecosystem succession. *Annual Review of Ecology and Systematics* 10:53-84.

Greenwood, D.J, Gastal, F., Lemaire, G., Draycott, A., Millard, P. & Neeteson, J.J. 1991. Growth rate and %N of field grown crops: Theory and experiment. *Annals of Botany* 67:181-190.

Grime, J.P. 1979. *Plant Strategies and Vegetation Processes*, John Wiley, Chichester.

Grubb P.J. 1989. The role of mineral nutrients in the tropics: a plant ecologist's view. In Proctor, J. (ed.) *Mineral Nutrients in Tropical Forest and Savanna Ecosystems*. Blackwell Scientific Publications, Oxford. pp. 417-439.

Haken, H. 1977. *Synergetics. An Introduction*. Springer-Verlag, Berlin.

Handbook of Chemistry and Physics. 1975. CRC Press, Cleveland.

Hart, S.C., Nason, G.E., Myrold, D.D. & Perry, D.A. 1994. Dynamics of gross nitrogen transformations in an old-growth forest: The carbon connection. *Ecology* 75:880-891.

He, X.-T., Stevenson, F.J., Mulvaney, R.L. & Kelley, K.R. 1988. Incorporation of newly immobilized ^{15}N into stable organic forms in the soil. *Soil Biology and Biochemistry* 20:75-81.

Heal, O.W. 1979. Decomposition and nutrient release in even-aged plantations. In: Ford, E.D. & Malcolm, D.C. (eds.) *The Ecology of Even-Aged Forest Plantations*. Institute of Terrestrial Ecology, Natural Environment Research Council, United Kingdom. pp. 257-291.

Hempfling, R., Ziegler, F., Zech, W. & Schulten, H.R. 1987. Litter decomposition and humification in acidic forest soils studied by chemical degradation, IR and NMR spectroscopy and pyrolysis field ionization mass spectrometry. *Zeitschrift für Pflanzenernährung und Bodenkunde* 150:179-186.

Hilbert, D.W. 1990. Optimization of plant root:shoot ratios and internal nitrogen concentration. *Annals of Botany* 66:91-99.

Howard, P.J.A. & Howard, D.M. 1974. Microbial decomposition of tree and shrub leaf litter. I. Weight loss and chemical composition of decomposing litter. *Oikos* 25:341-352.

Hunt, R. & Nicholls, A.O. 1986. Stress and the coarse control of growth and root-shoot partitioning in herbaceous plants. *Oikos* 47:149-158.

Hunt, H.W., Stewart, J.W.B. & Cole, C.V. 1983. A conceptual model for interactions among carbon, nitrogen, sulfur and phosphorus in grasslands. In: Bolin, B. & Cook, R.B. (eds.) *The Major Biogeochemical Cycles and Their Interactions*. SCOPE 21. John Wiley & Sons, Chichester. pp. 303-325.

Hunter, I.R., Nicholson, G. & Thorn, A.J. 1985. Chemical analysis of pine litter: An alternative to foliage analysis. *New Zealand Journal of Forestry Science* 15:101-110.

Hyvönen, R., Ågren, G.I. & Andrén, O. 1996. Modelling long-term carbon and nitrogen dynamics in an arable soil receiving organic matter and nitrogen amendments. *Ecological Applications* 6:1345-1354.

Ingestad, T. 1979. Nitrogen stress in birch seedlings. II. N, K, P, Ca, and Mg nutrition. *Physiologia Plantarum* 45:149-157.

Ingestad, T. 1980. Growth, nutrition, and nitrogen fixation in grey alder at varied rate of nitrogen addition. *Physiologia Plantarum* 50:353-364.

Ingestad, T. 1981. Nutrition and growth of birch and grey alder seedlings in low conductivity solutions and at varied relative rates of nutrient addition. *Physiologia Plantarum* 52:454-466.

Ingestad, T. 1982. Relative addition rate and external concentration; Driving variables used in plant nutrition research. *Plant, Cell and Environment* 5:443-453.

Ingestad, T. & Ågren, G.I. 1984. Fertilization for long-term maximum production. In: Perttu, K. (Ed.) *Ecology and Management of Forest Biomass Production Systems*. Department of Ecology & Environmental Research, Swedish University of Agricultural Sciences, Report 15. pp. 155-165.

Ingestad, T. & Ågren, G.I. 1988. Nutrient uptake and allocation at steady-state nutrition. *Physiologia Plantarum* 72:450-459.

Ingestad, T. & Ågren, G.I. 1991. The influence of plant nutrition on biomass allocation. *Ecological Applications* 1:168-174.

Ingestad, T. & Ågren, G.I. 1992. Plant nutrition and growth. *Physiologia Plantarum* 84:177-184.

Ingestad, T. & Ågren, G.I. 1995. Plant nutrition and growth: Basic principles. *Plant and Soil* 168/169:15-20.

Ingestad, T. & Kähr, M. 1985. Nutrition and growth of coniferous seedlings at varied relative nitrogen addition rate. *Physiologia Plantarum* 65:109-116.

Ingestad, T. & Lund, A.B. 1979. Nitrogen stress in birch seedlings. I. Growth technique and growth. *Physiologia Plantarum* 45:137-148.

Ingestad, T. & McDonald, A.J.S. 1989. Interaction between nitrogen and photon flux density in birch seedlings at steady-state nutrition. *Physiologia Plantarum* 77:1-11.

Ingestad, T., Aronsson, A. & Ågren, G.I. 1981. Nutrient flux density model of mineral nutrition in conifer ecosystems. *Studia Forestalia Suecia* 161:61-72.

Ingestad, T., Hellgren, O. & Ingestad Lund, A.B. 1994a. Data base for tomato plants at steady state. Methods and performance of tomato plants (*Lycopersicon esculentum* Mill. cv. solentos) under non-limiting conditions and under limitation by nitrogen and light. *Department of Ecology and Environmental Research. Swedish University of Agricultural Sciences. Report 74.* 50 pp.

Ingestad, T., Hellgren, O. & Ingestad Lund, A.B. 1994b. Data base for birch plants at steady state. Performance of birch plants (*Betula pendula* Roth.) under non-limiting conditions and under limitation by nitrogen and light. *Department of Ecology and Environmental Research. Swedish University of Agricultural Sciences. Report 75.* 38 pp.

Insam, H. & Domsch, K.H. 1988. Relationship between soil organic carbon and microbial biomass on chronosequences of reclamation sites. *Microbial Ecology* 15:177-188.

Insam, H. Parkinson, D. & Domsch, K.H. 1989. Influence of macroclimate on soil microbial biomass. *Soil Biology and Biochemistry* 21:211-221.

Jansson, S.L. 1958. Tracer studies on nitrogen transformations in soil with special attention to mineralization-immobilization relationships. *Annals of the Royal Agricultural College Sweden* 24:101-361.

Jarvis, P.G. & Sandford, A.P. 1986. Temperate forests. In: Baker, N.R. & Long, S.P. (eds.) *Photosynthesis in Contrasting Environments.* Elsevier Science Publishers. pp. 199-236.

Jenkinson, D.S. & Rayner, J.H. 1977. The turnover of soil organic matter in some of the Rothamsted classical experiments. *Soil Science* 123:298-305.

Jenkinson, D.S., Fox, R.H. & Rayner, J.H. 1985. Interactions between fertilizer nitrogen and soil nitrogen - the so-called -'priming' effect. *Journal of Soil Science* 36:425-444.

Jia, H. & Ingestad, T. 1984. Nutrient requirements and stress response of Populus simonii and Paulownia tomentosa. *Physiologia Plantarum* 62:117-124.

Johansson, M.B. 1995. The chemical composition of needle and leaf litter from Scots pine, Norway spruce and white birch in Scandinavian forests. *Forestry* 68:49-62.

Johnson, D.W 1992. Nitrogen retention in forest soils. *Journal of Environmental Quality* 21:1-12.

Johnson, I.R. & Thornley, J.H.M. 1987. A model of shoot:root partitioning with optimal growth. *Annals of Botany* 60:133-142.

Kaila, A. 1949. Biological absorption of phosphorus. *Soil Science* 68:279-289.

Kelley, K.R. & Stevenson, F.J. 1987. Effects of carbon source on immobilization and chemical distribution of fertilizer nitrogen in soil. *Soil Science Society of America Journal* 51:946-951.

Kirchmann, H., Persson, J. & Carlgren, K. 1994. The Ultuna long-term soil organic matter experiment, 1956-1991. *Department of Soil Sciences, Swedish University of Agricultural Sciences, Reports and Dissertations 17*, 55 pp.

Kristensen, E. 1990. Characterization of biogenic organic matter by stepwise thermogravimetry (STG). *Biogeochemistry* 9:135-159.

Ladd, J.N., Oades, J.M. & Amato, M. 1981. Microbial biomass formed from 14C, 15N-labelled plant material decomposing in soil in the field. *Soil Biology and Biochemistry* 13:119-126.

Landsberg, J.J. 1986. *Physiological Ecology of Forest Production.* Academic Press, London.

Legendre, L. & Legendre, P. 1983. *Numerical Ecology. Developments in Environmental Modelling*, 3. Elsevier, Amsterdam.

Lekkerkerk, L., Lundkvist, H., Ågren, G.I., Ekbohm, G. & Bosatta, E. 1990. Decomposition of heterogeneous substrates: An experimental investigation of a hypothesis on substrate and microbial properties. *Soil Biology and Biochemistry* 22:161-167.

Levi, M.P. & Cowling, E.B. 1969. Role of nitrogen in wood deterioration. VII. Physiological adaptation of wood-destroying and other fungi to substrates deficient in nitrogen. *Phytopathology* 59:460.468.

Levin, S.A., Mooney, H.A. & Field, C. 1989. The dependence of plant shoot:root ratios on internal nitrogen concentration. *Annals of Botany* 64:71-75.

Liebig, J.F. von 1840. *Chemistry and its Application to Agriculture and Physiology.* Taylor & Walton, London.

Likens, G.E. 1992. *The Ecosystem Approach: Its Use and Abuse.* Ecology Institute, Oldenburg/Luhe, Germany.

Linder, S. 1985. Potential and actual production in Australian forest stands. In: Landsberg, J.J. & Parsons, W. (eds.) *Research for Forest Management.* CSIRO, Melbourne. pp. 11-35.

Linder, S. & Rook, D.A. 1984. Effects of mineral nutrition on carbon dioxide exchange and partitioning in trees. In: Bowen, G.D. & Nambiar, E.K.S. (eds.) *Nutrition of Plantation Forests.* Academic Press, London. pp. 211-236.

Lohm, U., Larsson, K. & Nömmik, H. 1984. Acidification and liming of coniferous forest soil: long-term effects on turnover rates of carbon and nitrogen during an incubation experiment. *Soil Biology and Biochemistry* 16:343-346.

Mälkönen E. & Kukkola, M. 1991. Effect of long-term fertilization on the biomass production and nutrient status of Scots pine stands. *Fertilizer Research* 27:113-127.

Mattsson, M., Johansson, E., Lundborg, T., Larsson, M. & Larsson, C.M. 1991. Nitrogen utilization in N-limited barley during vegetative and generative growth. I. Growth and nitrate uptake kinetics in vegetative cultures grown at different relative addition rates of nitrate-N. *Journal of Experimental Botany* 42:197-205.

May, R.M. 1976. Simple mathematical models with very complicated dynamics. *Nature* 261:459-467.

McClaugherty, C. & Berg, B. 1987. Cellulose, lignin and nitrogen concentrations as rate regulating factors in late stages of forest litter decomposition. *Pedobiologia* 30:101-112.

McClaugherty, C.A., Pastor, J., Aber, J.D. & Melillo, J.M. 1985. Forest litter decomposition in relation to soil nitrogen dynamics and litter quality. *Ecology* 66:266-275.

Macdowall, F.D.H. 1972. Growth kinetics of Marquis wheat. I. Light dependence. *Canadian Journal of Botany* 50:89-99.

McMurtrie, R.E., Leuning, R., Thompson, W.A. & Wheeler, A.M. 1992a. A model of canopy photosynthesis and water use incorporating a mechanistic formulation of leaf CO_2 exchange. *Forest Ecology and Management* 52:261-278.

McMurtrie, R.E., Comins, H.N., Kirschbaum, M.U.F. & Wang, Y.-P. 1992b. Modifying existing forest growth models to take account of effects of elevated CO_2. *Australian Journal of Botany* 40:657-678.

McMurtrie, R.E., Gholz, H.L., Linder, S. & Gower, S.T. 1994. Climatic factors controlling the productivity of pine stands: a model-based analysis. *Ecological Bulletins* 43:173-188.

Melillo, J.M., Aber, J.D. & Muratore, J.F. 1982. Nitrogen and lignin control of hardwood leaf litter decomposition dynamics. *Ecology* 63:621-626.

Melin, J. 1986. Turnover and distribution of fertiliser nitrogen in three coniferous ecosystems in central Sweden. *Report No. 55. Department of Forest Soils. Swedish University of Agricultural Sciences.*

Melin, J. & Nömmik, H. 1988. Fertilizer nitrogen distribution in a Pinus sylvestris/Picea abies ecosystem, Central Sweden. *Scandinvian Journal of Forest Research* 3:3-15.

Melin, J., Nömmik, H., Lohm, U. & Flower-Ellis, J. 1983. Fertilizer nitrogen budget in a Scots pine ecosystem attained using root-isolated plots and 15N tracer technique. *Plant and Soil* 74:249-263.

Mengel, K. & Kirkby, E.A. 1979. *Principles of Plant Nutrition.* International Potash Institute, Berne, Switzerland.

Minderman, G. 1968. Addition, decomposition and accumulation of organic matter in forests. *Journal of Ecology* 56:355-362.

Monteith, J.L. 1977. Climate and efficiency of crop production in Britain. *Philosophical Transactions of the Royal Society* London Series B 281:277-294.

Mortland, M.M. & Wolcott, A.R. 1965. Sorption of inorganic nitrogen compounds by soil materials. In: Bartholomew, W.V. & Clark, F.E. (eds.) *Soil Nitrogen.* Agronomy 10. Amer. Soc. Agron., Madison, Wisconsin. pp. 150-197.

Niemann, G.J., Pureveen, J.B.M., Eijkel, G.B., Poorter, H. & Boon, J.J. 1992. Differences in relative growth rate in 11 grasses correlate with differences in chemical composition as determined by pyrolysis mass spectrometry. *Oecologia* 89:567-573.

Nömmik, H. 1978. Mineralization of carbon and nitrogen in forest humus as influenced by additions of phosphate and lime. *Acta Agriculturae Scandinavica* 28:221-230.

Nömmik, H. & Larsson, K. 1989. Assessment of fertiliser nitrogen accumulation in Pinus sylvestris trees and retention in soil by ^{15}N recovery technique. *Scandinavian Journal of Forest Research* 4:427-442.

Nömmik, H. & Nilsson, K.O. 1963. Fixed ammonia by the organic fraction of the soil. *Acta Agriculturae Scandinavica* 13:371-390.

Nömmik, H., Larsson, K. & Lohm, U. 1984. *Effects of experimental acidification and liming on the transformation of carbon, nitrogen and sulphur in forest soils.* National Swedish Environmental Protection Board Report 1869, Stockholm.

Nordén, B. & Berg, B. 1990. A non-destructive method (solid state ^{13}C NMR) for determining organic chemical components of decomposing litter. *Soil Biology and Biochemistry* 22:271-275.

Nyborg, M. & Hoyt, P.B. 1978. Effects of soil acidity and liming on mineralization of soil nitrogen. *Canadian Journal of Soil Science* 58:331-338.

Nye, P.H. & Tinker, P.B. 1977. *Solute Movement in the Soil-Root System. Studies in Ecology*, Vol. 4. Blackwell, Oxford.

Odum, E.P. 1971a. *Fundamentals of Ecology.* W.B. Saunders Co., Philadelphia, Pa.

Odum, H.T. 1971b. *Environment, Power and Society.* John Wiley, New York.

Odum, H.T. 1983. *Systems Ecology: An Introduction.* John Wiley, New York.

Olsen, C. 1950. The significance of concentration for the rate of ion absorption by higher plants in water culture. *Physiologia Plantarum* 3:152-164.

Olson, J.S. 1963. Energy storage and the balance of producers and decomposers in ecological systems. *Ecology* 44:322-331.

O'Neill, R.V., DeAngelis, D.L., Waide, J.B. & Allen, T.F.H. 1986. *A Hierarchical Concept of Ecosystems.* (Monographs in Population Biology; 23). Princeton University Press, Princeton.

Parnas, H. 1975. Model for decomposition of organic material by microorganisms. *Soil Biology and Biochemistry* 7:161-169.

Parton, W.J., Schimel, D.S., Cole, C.V. & Ojima, D.S. 1987. Analysis of factors controlling soil organic matter levels in Great Plains grasslands. *Soil Science Society of America Journal* 51:1173-1179.

Parton, W.J., Stewart, J.W.B. & Cole, C.V. 1988. Dynamics of C, N, P and S in grassland soils: A model. *Biogeochemistry* 5:109-131.

Pastor, J. & Post, W.M. 1986. Influence of climate, soil moisture, and succession on forest carbon and nitrogen cycles. *Biogeochemistry* 2:3-27.

Paustian, K. & Schnürer, J. 1987. Fungal growth response to carbon and nitrogen limitation: A theoretical model. *Soil Biology and Biochemistry,* 19:613-620.

Paustian, K., Ågren, G.I. & Bosatta, E. 1996. Modelling the role of litter quality on decomposition and nutrient cycling. In: Cadisch, G. & Giller, K.E. (eds.) *Driven by Nature: Plant Litter Quality and Decomposition* CAB International, Wallingford pp. 313-335.

Pearson, R.W. & Simonson, R.W. 1939. Organic phosphorus in seven Iowa soil profiles: Distribution and amounts as compared to organic carbon and nitrogen. *Soil Science Society of America Proceedings* 4:162-167.

Penning de Vries, F.W.T. 1974. Substrate utilization and respiration in relation to growth and maintenance in higher plants. *Netherlands Journal of Agricultural Science* 22:40-44.

Persson, J. 1980. Detailed investigations of the soil organic matter in a long term frame trial. *Department of Soil Sciences, Swedish University of Agricultural Sciences, Report* 128.

Persson, T. & Wirén, A. 1989. Microbial activity in forest soils in relation to acid/base and carbon/nitrogen status. *Meddelser fra Norsk Institutt for Skogsforskning* 42:83-94.

Persson, T., Wirén, A. & Andersson, S. 1991. Effects of liming on carbon and nitrogen mineralization in coniferous forests. *Water, Air and Soil Pollution* 54:351-364.

Peterjohn, W.T., Melillo, J.M., Steudler, P.A., Newkirk, K.M., Bowles, F.P. & Aber, J.D. 1994. Responses of trace gas fluxes and N availability to experimentally elevated soil temperatures. *Ecological Applications* 4:617-625.

Peters, R.H. 1991. *A Critique for Ecology.* Cambridge University Press, Cambridge.

Pettersson, R. & McDonald, A. J. S. 1994. Effects of nitrogen supply on the acclimation of photosynthesis to elevated CO_2. *Photosynthesis Research* 39:389-400.

Pettersson, R., McDonald, A. J. S. & Stadenberg, I. 1993. Response of small birch plants (*Betula pendula* Roth.) to elevated CO_2 and nitrogen supply. *Plant, Cell and Environment* 16:1115-1121.

Pierre, W.H. & Parker, F.W. 1927. Soil phosphorus studies: II. The concentration of organic and inorganic phosphorus in the soil solution and soil extracts and the availability of organic phosphorus to plants. *Soil Science* 24:119-128.

Piper, C.S. 1942. Investigations on copper deficiency in plants. *Journal of Agricultural Science* 42:143-178.

Poorter, H. & Bergkotte, M. 1992. Chemical composition of 24 wild species differing in relative growth rate. *Plant, Cell and Environment* 15:221-229.

Post, W.M., Pastor, J., Zinke, P.J. & Stangenberger, A.G. 1985. Global patterns of soil nitrogen storage. *Nature* 317:613-616.

Powlson, D.S., Brookes, P.C. & Christensen, B.T. 1987. Measurement of soil microbial biomass provides an early indicator of changes in total soil organic matter due to straw incorporation. S*oil Biology and Biochemistry* 19:159-164.

Rastetter, E.B. & Shaver, G.R. 1992. A model of multiple-element limitations for acclimating vegetation. *Ecology* 73:1157-1174.

Rastetter, E.B., Ryan, M.G. Shaver, G.R., Melillo, J.M., Nadelhoffer, K.J., Hobbie, J.E. & Aber, J.D. 1991. A general biogeochemical model describing the response of the C and N cycles in terrestrial ecosystems to changes in CO_2, climate, and N deposition. *Tree Physiology* 9:101-126.

Rastetter, E.B., King, A.W., Cosby, B.J., Hornberger, G.M., O'Neill, R.V. & Hobbie, J.E. 1992. Aggregating fine-scale ecological knowledge to model coarser-scale attributes of ecosystems. *Ecological Applications* 2:55-70.

Rastetter, E.B., Ågren, G.I. & Shaver, G.R. 1997. Responses to increased CO_2 concentration in N-limited ecosystems: Application of a balanced-nutrition, coupled-element-cycling model. *Ecological Applications* 7:444-460.

Reddy, K.R. & Patrick,W.H. 1978. Residual fertilizer nitrogen in a flooded rice soil. *Soil Science Society of America Journal* 42:316-318.

Redfield, A.C. 1958. The biological control of chemical factors in the environment. *American Scientist* 46:205-221.

Richards, E.H. & Norman, A.G. 1931. The biological decomposition of plant material. V. Some factors determining the quantity of nitrogen immobilized during decomposition. *Biochemistry Journal* 25:1769-1778.

Rodin, L.E. & Basilevich, N.I. 1967. *Production and Mineral Cycling in Terrestrial Ecosystems*. Oliver and Boyd, Edinburgh.

Rolff, C. & Ågren, G.I. 199X. A model study of nitrogen limited forest growth (Unpublished manuscript).

Ryan, M.G. 1991. Effects of climate change on plant respiration. *Ecological Applications* 1:157-167.

Ryan, M.G., Aber, J.D., Ågren, G.I., Friend, A.D., Linder, S., McMurtrie, R.E., Parton, W.J., Raison, R.J. & Rastetter, E.B. 1996a. Comparing models of ecosystem function for temperate conifer forests. I. Model description and validation. In: Breymeyer, A., Hall, D.O., Melillo, J.M. & Ågren, G.I. (eds.) *Global Change: Effects on Forests and Grasslands*, J.Wiley, Chichester pp. 313-362.

Ryan, M.G., Aber, J.D., Ågren, G.I., Friend, A.D., McMurtrie, R.E., Parton, W.J.& Rastetter, E.B. 1996b. Comparing models of ecosystem function for temperate conifer forests. II. Predictions. In: Breymeyer, A., Hall, D.O., Melillo, J.M. & Ågren, G.I. (eds.) *Global Change: Effects on Forests and Grasslands*, J.Wiley, Chichester pp. 363-387.

Saggar, S., Bettany, J.R. & Stewart, J.W.B. 1981. Measurement of microbial sulfur in soil. *Soil Biology and Biochemistry* 13:493-498.

Schimel, J.P. & Firestone, M.K. 1989. Inorganic N incorporation by coniferous forest floor material. *Soil Biology and Biochemistry* 21:41-46.

Schlesinger, W.H. 1991. *Biogeochemistry: An Analysis of Global Change*. Academic Press, San Diego.

Schollenberg, C.J. 1920. Organic phosphorus content of Ohio soils. *Soil Science* 10:127-141.

Scott, N.M. 1985. Sulphur in soils and plants. In: Vaughan, D. & Malcolm, R.E. (eds.) *Soil Organic Matter and Biological Activity. Developments in Plant and Soil Sciences* Vol. 16. Martinus Nijhoff/Dr. W. Junk Publishers, Dordrecht. pp. 379-401.

Šesták, Z. (ed.) 1985. *Photosynthesis during leaf development*. Dr.W. Junk Publishers, Dordrecht.

Shaver, G.R. & Aber, J.D. 1996. Carbon and nutrient allocation in terrestrial ecosystems. In: Breymeyer, A., Hall, D.O., Melillo, J.M. & Ågren, G.I. (eds.) *Global Change: Effects on Forests and Grasslands*, J.Wiley, Chichester pp. 183-198.

Shugart, H.H. 1984. *A Theory of Forest Dynamics*. Springer-Verlag, New York.

Sillén, L.G. 1966. Regulation of O_2, N_2 and CO_2 in the atmosphere; thoughts of a laboratory chemist. *Tellus* 18:198-206.

Smolders, E., Buysse, J. & Merckx, R. 1993. Growth of soil-grown spinach plants at different N-regimes. *Plant and Soil* 154:73-80.

Sollins, P., Grier, C.C., McCorison, F.M., Cromack, K. Jr., Fogel, R. & Fredriksen, R.L. 1980. The internal element cycles of an old-growth Douglas-fir ecosystem in western Oregon. *Ecological Monographs* 50:261-285.

Sparling, G. 1992. Ratio of microbial biomass carbon to soil organic carbon as a sensitive indicator of changes in soil organic matter. *Australian Journal of Soil Research* 30:195-207.

Stanford, G. & Smith, S.J. 1972. Nitrogen mineralization potentials of soils. *Soil Science Society of America, Proceedings* 36:465-472.

Steenbjerg, F. 1951. Yield curves and chemical plant analysis. *Plant and Soil* 3:97-109.

Stevenson, F.J. 1982. *Humus Chemistry. Genesis, Composition, Reactions.* Wiley, New York.

Stewart, J.W.B. 1984. Interrelation of carbon, nitrogen, sulfur and phosphorus cycles during decomposition processes in soil. In: Klug, M.J. & Reddy, C.A. (eds.) *Current Perspectives in Microbial Ecology.* American Society of Microbiology, Washington, D.C. pp. 442-446.

Swift, M.J., Heal, O.W. & Anderson, J.M. 1979. *Decomposition in Terrestrial Ecosystems.* (Studies in Ecology, Vol. 5). Blackwell, Oxford.

Tabatabai, M.A. & Al-Khafaji, A.A. 1980. Comparison of nitrogen and sulphur mineralization of microbial organic phosphorus in soil materials. *Soil Science Society of America Journal* 44:1000-1006.

Tamm, C.O. 1963. Upptagning av växtnäring efter gödsling av gran- och tallbestånd. *Inst. för skogsekologi. Skogshögskolan. Rapporter och Uppsatser* Nr 1.

Taylor, B.R.D., Parkinson, D. & Parsons, W.F.J. 1989. Nitrogen and lignin content as predictors of litter decay rates: a microcosm experiment. *Ecology* 70:97-104.

Thompson, L.M. & Black, C.A. 1950. The mineralization of organic phosphorus, nitrogen, and carbon in Clarion and Webster soils. *Soil Science Society of America Proceedings* 14:147-151.

Thompson, L.M., Black, C.A. & Clark, F.E. 1948. Accumulation and mineralization of microbial organic phosphorus in soil materials. *Soil Science Society of America Proceedings* 13:242-245.

Thompson, L.M., Black, C.A. & Zoellner, J.A. 1954. Occurence and mineralization of organic phosphorus in soils, with particular reference to associations to nitrogen, carbon, and pH. *Soil Science* 77:185-196.

Thornley, J.M.T. 1976. *Mathematical Models in Plant Physiology*, Academic Press, London.

Tilman, D. 1988. *Plant Strategies and the Dynamics and Structure of Plant Communities*. Monographs in Population Biology 26. Princeton University Press, Princeton, N.J.

van Oene, H. 1992. Acid deposition and forest nutrient imbalances: A modelling approach. *Water, Air, and Soil Pollution* 63:33-50.

van Oene, H. & Ågren, G.I. 1995a. The application of the NAP model to the Solling spruce site. *Ecological Modelling* 83:139-149.

van Oene, H. & Ågren, G.I. 1995b. Complexity and simplicity in modelling of acid deposition effects on forest growth. *Ecological Bulletins* 44:352-362.

van Veen, J.A. & Paul, E.A. 1981. Organic carbon dynamics in grasslands soils. I. Background information and computer simulations. *Canadian Journal of Soil Science* 61:185-201.

Vitousek, P. 1982. Nutrient cycling and nutrient use efficiency. *American Naturalist* 119:553-572.

Waksman, S.A. 1924. Influence of microorganisms upon the carbon:nitrogen ratio in the soil. *Journal of Agricultural Science* 14:555-562.

Waksman, S.A. & Tenney, F.G. 1927. The composition of natural organic materials and their decomposition in soil. II. Influence of age of plant upon the rapidity and nature of its decomposition - rye plants. *Soil Science* 24:317-334.

Walker, T.W. & Adams, A.F.R. 1958. Studies on soil organic matter: I. Influence of phosphorus content of parent material on accumulation of carbon, nitrogen, sulphur, and organic phosphorus in grassland soils. *Soil Science* 85:307-318.

Wessén, B. & Berg, B. 1986. Long-term decomposition of barley straw: chemical changes and ingrowth of fungal mycelium. *Soil Biology and Biochemistry* 18:53-59.

White, C.S., Moore, D.E., Horner, J.D. & Gosz, J.R. 1988. Nitrogen mineralization-immobilization response to field N or C perturbations: an evaluation of a theoretical model. *Soil Biology and Biochemistry* 20:101-105.

White, D.L., Haines, B.L. & Boring, L.R. 1988. Litter decomposition in southern Appalachian black locust and pine-hardwood stands: litter quality and nitrogen dynamics. *Canadian Journal of Forest Research* 18:54-63.

Wikström, J. F. 1994. A theoretical explanation of the Piper-Steenbjerg effect. *Plant, Cell and Environment* 17: 1053-1060.

Wikström, F. & Ågren, G. I. 1995. The relationship between the growth rate of young plants and their total-N concentration is unique and simple. *Annals of Botany* 75:541-544.

Williams, K., Percival, F., Merino, J. & Mooney, H.A. 1987. Estimation of tissue construction cost from heat of combustion and organic nitrogen content. *Plant, Cell and Environment* 10:725-724.

Wolters, V. & Joergensen, R.G. 1991. Microbial carbon turnover in beech forest soils at different stages of acidification. *Soil Biology and Biochemistry* 23:897-902.

Zelawksi, W., Lotocki, A., Morteczka, H., Pryzkorska-Zelawski, T. & Wrzesniewski, W. 1985. Growth analysis of Scots pine (*Pinus sylvestris* L.) seedlings cultivated in a wide range of experimental conditions. *Acta Societas Botanicorum Poloniae* 54:255-272.

Zheng, D.W., Bengtsson, J. & Ågren, G.I. 1997. Soil food webs and ecosystem processes: Decomposition in donor-control and Lotka-Volterra systems. *American Naturalist* 149:125-148.

Solutions to selected problems

Problem 3.8 Assuming a constant rate of litter input and combining (3.1) with (3.33) gives

$$R = f_C u \frac{1-e_0}{e_0} \frac{I_0}{k}\left(1-e^{-kt}\right) = I_0\left(1-e^{-kt}\right) \to I_0$$

Problem 3.10 From (3.31) we get

$$N_s^{SS} = I_0 \int_0^\infty h_N(a)da = I_0 \int_0^\infty \left[\frac{f_N}{f_C} - \left(\frac{f_N}{f_C} - r_0\right)e^{-f_C u a}\right]e^{-(1-e_0)f_C u a/e_0}da =$$

$$= \frac{I_0}{k}\left[r_c + r_0(1-e_0)\right]$$

Hence, $r^{SS} = r_c + r_0(1-e_0) > r_c$ and the soil is C limited.

Problem 4.1 Priming effects means that the availability of high quality substrates increases the decomposition rate of low quality substrates. This means that the assimilation of a given quality depends, not only on how much of that quality is present, but also on how much is present of all other qualities. This could be described by a function $u_p(q)$ that takes into account all qualities, e.g.

$$u_p(q) = \int \omega(q,q'')\rho_C(q'',t)dq''$$

$u(q)$ in (4.7) should the be replaced by $u(q) + u_p(q)$.

Problem 4.3 $D(q,q')$ does not appear explicitly in (4.9) or (4.10). However, the shape of ρ_C depends on D and through it $\hat{q}$ and hence $k(\hat{q})$.

Problem 4.4 (4.9) is an exact expression for the time derivative of C. All the terms in the integrand are positive. Hence, $dC/dt < 0$.

Problem 4.5 Inserting ρ_C and D in (4.7) gives

$$\frac{\partial}{\partial t}\sum_i C_i(t)\delta(q-q_i) = \sum_i \frac{dC_i(t)}{dt}\delta(q-q_i) = -f_C\sum_i \frac{1-e(q)}{e(q)}u(q)C_i(t)\delta(q-q_i) +$$

$$+ f_C\int\left[\sum_j d(q,q')\delta(q-q_j)u(q')\sum_i C_i(t)\delta(q'-q_i)\right]dq' =$$

$$= -f_C\sum_i \frac{1-e_i}{e_i}u_i C_i(t)\delta(q-q_i) + f_C\sum_{i,j} d(q,q_j)u_j C_i(t)\delta(q-q_i)$$

Integration over intervals that only contain one q_i then gives a system of equations.

Problem 4.6 From (4.13) we get the following expression for the total amount of carbon

$$C(t) = \int\rho_C(q,0)e^{-k(q)t}dq = \int\sum_i C_i\delta(q-q_i)e^{-k(q)t}dq =$$

$$= \sum_i C_i e^{-k(q_i)t}$$

Problem 4.8 Integration of (4.17) gives

$$q_t = q_0 e^{-f_C\eta_{10}u_0 t} \xrightarrow[t\to\infty]{} 0$$

and integration of (4.18)

$$C(t) = C_0 e^{-\frac{1-e_0}{\eta_{10}e_0}(q_0-q_t)} \xrightarrow[t\to\infty]{} C_0 e^{-\frac{1-e_0}{\eta_{10}e_0}q_0}$$

and from (4.10)

$$k(t) = f_C\frac{1-e_0}{e_0}u_0 q_0 e^{-f_C\eta_{10}u_0 t} \xrightarrow[t\to\infty]{} 0$$

Problem 4.9 The addition of an extra term to D will add an extra term to (4.15)

$$f_C\int\eta_2\delta''(q-q')u(q')\rho_C(q',t)dq' = f_C\frac{\partial^2}{\partial q^2}\left[u(q)\rho_C(q,t)\right]$$

where we have used (A1.5)

Problem 4.10 The rate of change in quality is to first order given by $dq/dt = -f_C u \eta_1$. Since both f_C and u are positive, η_1 has also to be positive if dq/dt is negative.

Problem 4.13 From (4.35) we get the following implicit equation for q_c

$$r_{nc} = \frac{f_n}{f_C} e_1 q_c^\alpha = \frac{f_n}{f_C} - \left(\frac{f_n}{f_C} - r_0\right) e^{-\int_{q_c}^{q_0} \frac{dq}{\eta_1(q)}} = \frac{f_n}{f_C} - \left(\frac{f_n}{f_C} - r_0\right) e^{-\frac{q_0 - q_c}{\eta_{10}}}$$

Combined with the solution to Problem 4.8, this gives an expression for t_c.

Problem 4.14 In the first order approximation, q and t are uniquely linked and evolution in q is equivalent to temporal evolution. When higher order moments are included the direct link in the relationship between q and t depends also on these moments.

Problem 4.15 From (4.38) we get

$$g(q) = \exp\left\{\frac{1}{\eta_{10} e_1}\left[\frac{q^{1-\alpha} - q_0^{1-\alpha}}{1-\alpha} + e_1(q_0 - q)\right]\right\}$$

and $q(t)$ is given by (4.46). These functions can now be plotted for different parameter values ($e_1 q_0^\alpha < 1$).

Problem 4.17 In the limit of small q, $e(q) = e_0 + e_1 q \approx e_0$ when $e_0 \neq 0$ and the divergence of the integral in the solution to (4.36) is determined by the behaviour of $\eta_1(q)$

$$g(q) = \exp\left\{-\int_q^{q_0} \frac{1 - e(q')}{\eta_1(q')e(q')} dq'\right\}$$

This integral converges when $\eta_1(q)$ approaches 0 more slowly than q and $g(q) > 0$.

Problem 4.18 (4.47) is now

$$C(t) = \frac{I_0}{f_C \eta_{11} u_0 q_0^\beta} \int_{q_t}^{q_0} \left[\frac{\eta_{10} + \eta_{11} q}{\eta_{10} + \eta_{11} q_0}\right]^{\frac{1-e_0}{\eta_{11} e_0}} \frac{dq}{q^{\beta+1}}$$

The behaviour of this integral for small q is entirely determined by the term $q^{\beta+1}$ and because $\beta > 0$, it will diverge for all values of β, which corresponds to an infinite accumulation of carbon.

Problem 5.3 If $q_0 \cong q_c$ in the solution to problem 4.13, the exponential can be expanded to first order which gives the following equation

$$e_1 q_c = \frac{r_0}{f_N / f_C} + \left(1 - \frac{r_0}{f_N / f_C}\right) \frac{q_0 - q_c}{\eta_{10}}$$

from which we get

$$q_c = \frac{\eta_{10} \frac{r_0}{f_N / f_C} + \left(1 - \frac{r_0}{f_N / f_C}\right) q_0}{e_1 \eta_{10} + \left(1 - \frac{r_0}{f_N / f_C}\right)} \approx \frac{q_0}{1 + e_1 q_0} \approx q_0$$

where we have successively used $r_0/(f_N/f_C) \ll 1$ and $e_1 q_0 \ll 1$ (cf. Table 5.1).

Problem 7.3 The rate of mineralisation is given by (7.23). In this expression, the first terms with q_i are equal for both the perturbed and unperturbed system. If the time for equal rates occurs for $t \leq t_l$, we use the first line of (7.23). For the unperturbed system, the last two terms (inside the square brackets) are equal. The time for equal mineralisation is thus determined by the condition that these two terms are equal for the perturbed system. That the gs are equal is equivalent to the qs being equal. Combining (7.22) and (7.19) gives $(\alpha = \beta f_C \eta_{11} u_0 q_0{}^\beta)$

$$\left(\frac{q_0}{q_\ell}\right)^\beta \left(1 - \frac{u_0}{u_p}\right) + \frac{u_0}{u_p} + \alpha \frac{u_0}{u_p} t_\ell + \alpha t = 1 + \alpha \frac{u_0}{u_p} t$$

from which t can be calculated. If $t > t_l$, the two terms within the square brackets do not cancel and we are left with the following expression from which t can be calculated

$$\left[1+\alpha(t_1+t)\right]^x - \left[1+\alpha t\right]^x =$$

$$= \frac{u_p}{u_0}\left\{\left[\left(\frac{q_0}{q_p}\right)^\beta\left(1-\frac{u_0}{u_p}\right)+\frac{u_0}{u_p}+\alpha\left(\frac{u_0}{u_p}t_1+t\right)\right]^x - \left[\left(\frac{q_0}{q_p}\right)^\beta\left(1-\frac{u_0}{u_p}\right)+\frac{u_0}{u_p}+\alpha t\right]^x\right\}$$

where $x = -(1-e_0)/(\beta e_0 \eta_{11})$.

Problem 7.5 The border between mineralisation and immobilisation is that $m_C^P(0) = 0$. Applying this to (7.23) we have

$$-h_N(q_i(t_p,0)) + h_N(q_i(t_\ell,0)) = \frac{u_p}{u_0}\left[h_N(q_{ii}(t_p,0)) - h_N(q_{iii}(0,0))\right]$$

Since we are normally considering old soils, t_p is very large $q_i(t_p,0) = q_{ii}(t_p,0) = 0$. Moreover, $q_{ii}(t_p,0) = q_l$ and $q_{iii}(0,0) = q_0$. We then get

$$h_N(q_\ell) = -\frac{u_p}{u_0}h_N(q_0) = -\frac{u_p}{u_0}r_{0N}$$

Problem 7.6 In section 7.1 we showed that net mineralisation is given by $f_C u N / e - f_N u C$ and that gross mineralisation is given by $f_C u N / e$. The ratio of net to gross mineralisation is thus

$$1-\frac{e_0 f_N C}{f_N N} = 1 - \frac{e_0}{1-\left(1-\dfrac{r_{0N} f_C}{f_N}\right)\left(1-\dfrac{e_0}{1-\beta\eta_{10}e_0}\right)}$$

where we have used (4.56) for the C/N ratio.

Problem 7.7 From (7.9) we have the equation for the inorganic part of the fertiliser

$$\frac{d\Delta_i}{dt} = -\frac{f_C}{e}u\Delta_i + f_N uC\frac{\Delta_i}{N_i+\Delta_i} \approx -\frac{f_C}{e}u\Delta_i \text{ if } \Delta_i/N_i \approx 0$$

Using (4.27) with $\sigma^2 = 0$, this equation can be converted into an equation in q

$$\frac{d\Delta_i}{dq} = \frac{1}{e\eta_1}\Delta_i$$

which using Model II has the solution

$$\Delta_i(t) = \Delta_i(t_0)\left[\frac{\hat{q}(t_0 + t)}{\hat{q}(t_0)}\right]^{1/n_{11}e_0}$$

Problem 8.2 If R_W is calculated as in (8.25), then c_N becomes also a function of P_N and k; $c_N = (P_N/k)^{1/2}$. When this is taken into account in (8.3), the two expressions become identical.

Problem 8.5 For small values of W ($W \ll K_2$), Greenwood's equation is $dW/dt \approx K_1 W/K_2$, which corresponds to exponential growth with a rate K_1/K_2 (8.34). For large values of W ($W \gg K_2$), the equation is $dW/dt \approx K_1$, which corresponds to linear growth rate (8.36). In the limits of small and large plants, (8.35) and (8.36) are therefore equal to Greenwood's equation. The difference is in the transition zone, where Greenwood's equation provides a smoother transition but at the expense of a somewhat more complicated expression.

Problem 8.6 (A) If the distribution is homogeneous such that each plant gets a nutrient supply corresponding to its size, the growth of the plants is described by (8.24) and the plants will grow exponentially with a rate $R = (P_n k)^{1/2}$. Of course, the supply is increasing exponentially in time to maintain a constant k. A small plant, starting with a size W_s will at time t have the size $W_s e^{Rt}$. A big plant, starting with a size W_b will at time t have the size $W_b e^{Rt}$. The relative sizes of the plants will, therefore not change over time but the absolute sizes, which is what counts, will increase exponentially.

(B) Consider now a situation where the nutrient additions give the plants equal amounts (and again, of course, expontially increasing).The amount of nutrient in a plant is then described by $n(t) = n_0 + n_1 e^{Rt}$ and the plant size becomes $W(t) = W_0 + P_n n_0 t + P_n n_1 (e^{Rt} - 1)/R$. Hence, with time the effects of the initial conditions are wiped out by the exponential term and all plants converge towards the same size.

Problem 9.1 Equation (9.5) becomes $dW/dt = (a - bW)N - fW$. At steady state we can get the following expression

$$W = \frac{a}{b} - \frac{f}{bc_N}$$

The mortality thus reduces the maximum set by the decreasing nitrogen productivity but this can be balanced by increasing nitrogen concentration.

Problem 11.1 One possibility to answer this problem is through a sensitivity analysis. Define the sensitivity of the total carbon store, C, to a parameter p as $100 \cdot d(\ln C)/d(\ln p)$. With Model b we get the following sensitivities:

Parameter	Value	Sensitivity
a	20	27
b	0.002	-110
μ_v	0.3	10
k_C	0.2	72
r_c	0.05	-52
r_0	0.02	-21
N_T	500	89

Subject index

abiotic reactions, 92, 97-8, 108
absorption function, 98
abstraction, 4
acid insoluble (-s), 186
acid soluble (-s), 186
added nitrogen interaction (ANI), 94, 96
affinity constant, 98-9
allocation models, 165
Alnus incana, 141
analytical solutions, 19, 39, 42, 167, 176
Appolonius, 5
approximation
  adiabatic, 9, 24
  alternative to exact solutions, 39
  first order, 30, 45, 48, 193
  zeroth order, 24, 30
aspen, 63, 65-6
assimilation rate, 14, 47, 148-50

barley, 73
beets, 73
Betula pendula, 132, 141, 152
BFG, 179-80, 185
BIOMASS model, 165
Black Hawk Island, 189
boundary layer, 14
box-transfer models, 38
Brahe, 5

cabbage, 141, 144
carbohydrates, 60
carbon concentration,
    decomposers, 20, 36
carbon dioxide concentration, 133
carbon mineralisation
  and acidification, 105

and liming, 92, 100ff
CENTURY model, 59, 165
chemical equilibrium, 12
  and steady state, 15
Classical Mechanics, 3-4, 7-8
climatic change, *see* global change
climatic factors, 42, 48, 52, 73, 108, 137, 146, 165, 167, 175, 187
cohort, litter, 19, 30ff, 47ff, 61-2, 68, 76-7, 80, 85-6, 91ff, 173, 177
computers, 5-6
concept, germinal, 3-4, 6
conceptual evolution, 4, 6

decomposer activity, 19, 100
decomposer biomass, 61
  and adiabatic approximation, 24
  change over time, 20ff, 35, 73ff
  in soil organic matter, 76
decomposer efficiency, 20, 34ff, 66-7, 85, 192
decomposer functions, models for, 44ff, 61ff, 107
decomposer growth rate, 20, 34ff
  in field conditions, 175
  in laboratory conditions, 85-7, 104
  perturbed by liming, 91
  time varying, 91, 107, 176
decomposer mortality rate, 21, 36-7, 48, 73, 76-7
decomposer production rate, 22, 25
decomposition subsystem, 32
deductive power, 5

density function
  dynamic equations, 34ff
  of carbon in decomposer
    biomass, 36
  of carbon in substrate, 34ff, 92
  of elements in substrate, 45ff
deposition, acid, 183
dimensional analysis, 114
Dirac delta function
  properties and definition, 195
  as model of dispersion process,
    39
discrete boxes models, 39
dispersion in quality, 34ff, 80, 85
  models for, 52, 57, 193, 196
dissolved organic carbon, 189
distribution of elements, 12-4, 153
DOC, *see* dissolved organic
    carbon
dose-response curves, 120, 123-4,
    126-7, 130

ecology
  ecosystem, 3, 6, 193
  evolutionary, 7
  subspaces, 7-8
ecosystem
  equation, 10, 159
  operator, 10, 160
  state vector, 10, 159
  steady state solution, 162
  stoichiometry, 161
  theory, 159
ecosystem ecology, *see* ecology,
    ecosystem
ecosystems and global change,
    167
Einstein, 4-5
element
  accumulation phase of, 20, 27-
    8, 32
  cycling, 3, 12ff, 35, 16

release phase of, 20-1, 28, 32
  stabilisation in soils, 79-80, 82ff
element:carbon ratio, 32, 62, 91,
    98, 101, 106
  of soil organic matter, 79, 85-6,
    161
element:carbon ratio, critical, 20-
    1, 23, 25, 32, 68, 80
  and quality, 46, 53
elemental distribution, 12ff
equations
  linear, 11, 38, 51, 60, 116, 119,
    120-1, 132, 193
  non-linear, 11, 80, 93, 193

Farquhar-von Caemmerer model,
    118, 138
fertilisation, 72, 105, 170
fertiliser, 92, 100-1, 107, 113, 170
forestry, short rotation, 174, 180
FORET model, 165

Galileo, 3-4
gamma function, 60
G′DAY model, 165
GEM model, 165
global change, 167, 179
global cycles, 14, 15
Grime model, 165

Hamilton, 3, 4
harvest, 73, 153, 175, 180-3
hemlock, 65
hierarchy, 10
humification, *see* humus
humus, 35, 40, 55

IBP, *see* International Biological
    Program
immobilisation potential of
    nitrogen fertiliser, 91, 93
incubation experiments, 79, 87-8

inorganic nitrogen, 20, 23, 26,
91ff, 161, 170
input, litter, 30-1, 54, 58, 70, 72,
76, 85, 88, 100-2, 106, 117,
170-1, 177, 187
International Biological Program,
113
isotope discrimination, 194

Kepler, 5

lag time, 24, 29-30
Lambert-Beer law, 144
language, 3-6, 10
leaching, nitrogen, 170, 176, 182,
184
leaf area, 14-5, 118-20, 137, 145
leaf maturation, 119-20
leaf weight ratio, 118-9
Lemna gibba, 141, 155
Lemna minor, 141, 155
Lemna paucicostata, 141, 155
lettuce, 141, 144
Liebig's law, 113, 131
light extinction, 139, 144-6
light intensity, 140, 142-6, 150,
155
light use efficiency, 137
lignin, 40, 60, 186
liming, 91
perturbation in C accessibility,
92, 100-1
effects on C and N
mineralisation, 92, 104
LINKAGES model, 165
log-normal distribution, 196
Lycopersicon esculentum, 141

Maclaurin series, 62, 130
magnesium, 183
manure
farmyard, 73
green, 73

mathematics, 3-6, 10
Medicago litoralis, 77
MEL model, 166
Mercury, 5
Michaelis-Menten, 97
mineralisation, acidification
effects, 105
Mitscherlich, 113
moment expansion, 42ff, 76
of the density function, 43
of the dispersion function, 43
and dynamic equations, 196
numerical solutions, 44, 196
mortality
decomposer, *see* decomposer
plant, 115, 148, 163

^{15}N, 91ff, 108
Newton, 3-4
nitrogen concentration, 15
decomposers, 20-1, 25, 92, 97,
173
litter, 62-3, 65, 67, 162, 171,
173, 190
plant, 114ff, 141ff, 171, 175,
185, 191, 194
nitrogen deposition, 170-4, 183
nitrogen factor, 20, 28
nitrogen immobilisation
abiotic, 97
biotic, 97, 108
effects of inorganic nitrogen,
91, 96
nitrogen limitation, 184
nitrogen, leaching phase of,
20-1
nitrogen mineralisation
and acidification, 105
and liming, 92, 104, 106
gross rate, 91, 107
in soil, 104-5, 161, 172-3
in soil profiles, 194
net rate, 23, 91-3, 107

nitrogen productivity, 15, 114ff
  models for, 139, 161
  and light, 140, 142ff
  and photosynthesis, 118
  and quality, 186, 191
nitrogen saturation, 170-1, 174
NMR, *see* nuclear magnetic
  resonance
non-observables, 4
non-polar soluble (-s), 186, 189
Norway spruce, *see Picea abies*
nuclear magnetic resonance, 192
nutrient
  budget for forest, 180
  excess, 170-1, 176, 183, 184
nutrient productivity, 131, 133,
  137-8, 160
  and species, 140
  experimental test, 139-40
nutrient use efficiency, NUE, 151-
  2

oats, 73
observable, 4, 186
open systems, 11-2
optimisation, 114, 150-1, 155
orthogonal factors, 131-2

Paulownia tomentosa, 141
peat, 61
Phanerochaete chrysosporium,
  61, 66
phosphorus, 13-4, 16, 34, 79ff,
  132
phosphorus mineralisation, 84-5,
  88, 90
photon flux density, 118
photosynthesis, 10, 113, 118-21,
  137-8, 142, 146, 148, 165
Picea abies, 66, 141, 146, 171,
  181-4
Pinus contorta, 141
Pinus nigra, 146

Pinus radiata, 179
Pinus resinosa, 146, 147
Pinus sylvestris, 52, 62-4, 68-72,
  132, 141, 146-7, 169, 171,
  176, 179, 187, 189-90
Piper-Steenbjerg effect, 151-2
plant carbon balance, 113, 118
plant growth, 14
  and nitrogen productivity, 15,
  115ff, 139ff
  anomalous relation with c_N, 152
  non-exponential, 133, 141
  response ranges, 116-7, 123,
  132, 137, 140
plant nutrient concentration
  and plant size, 125, 133, 152,
  154
  and photosynthetic rate, 119
  minimum, 115-6
  optimal, 116, 125, 133
Poria oleracea, 61, 66
production-to-assimilation ratio,
  20, 36
protection, physical, 52, 58, 80
protein, 15-6, 114
priming effect, 38
principal axis, 7, 9
Pseudotsuga menziesii, 146
Ptolemaeus, 5

Q_{10}, 87
qCO_2, 188
Q model, 177, 179-80, 183
q-representation, 48
quality, 9
  and chemical composition, 186,
  191
  and nitrogen productivity, 186,
  191
  average value, definition, 42,
  187
  bridge between plant and soil,
  186ff

of litters, 66ff, 101, 168, 173
of soil organic matter, 35
of substrates, 34ff, 100
perturbation in, 86, 92
quantum mechanics, 3-4, 6, 8, 39

rape, 73
red maple, 63, 65
reduced variables, 123, 125-7,
 129, 142
relative growth rate of plants
 and several nutrients, 132
 maximal, 116, 121, 139
 from optimising root:shoot
  ratios, 150-1, 155
relativistic mechanics, 4-5, 8
residence time, 51, 152
respiration rate
 of decomposers, 22, 32, 47,
  104-5, 198
 of plant, 118, 120, 138
  growth, 118
  maintenance, 118
retention capacity, 99-100
root absorbing power, 155
root:shoot ratios, 148ff
Rothamsted model, 59
Rubisco, 15, 118
ryegrass, 77

Salix viminalis, 146
de Saussure, 131
sawdust, 73
Schrödinger equation, 6
Scots pine, *see Pinus sylvestris*
senescence, 120
sewage sludge, 73
soil
 agricultural, 72, 76, 99-101, 188
 depth, 193-4
 forest, 98-101, 171-4
solutions, numerical, 5, 9, 44, 153,
 167, 176-7, 196-7, 199-200

specific decomposition rate, 26,
 29, 39-41, 49-50, 55, 59, 68-
 9, 168
 climatic effects, 187, 190
specific leaf area, 119
spectroscopy, 191-2
Sprengel, 131
steady state of soil organic matter,
 31, 61-2, 68-70
straw, 73, 77
substrate
 heterogeneous, 34, 37, 88, 177
 homogeneous, 19ff, 34, 40, 46,
  59, 67, 88, 91, 161
sugar, 40, 60
sulphur, 13-4, 16, 79, 82ff, 108
sulphur mineralisation, 85, 90
SWECON, 179, 180, 185

Taylor series, 24
temperature, 52, 76-7, 115, 131,
 175, 179-80, 187, 190
theory, 3
 as an infinite set of models, 6,
  10
 translation through a model, 6
thermodynamics, 11
Tilman model, 165
time constants, 8-10
tool, 3-4, 6, 10
toxicity, 117
turnover
 carbon, 14-5, 19, 70-1
 nitrogen, 14-5, 19, 70-1, 97,
  175

uptake of nutrients, 120ff
 exponential, 120, 122, 124ff,
  136
 linear, 122
 fixed amount, 122, 126ff

variance, 37, 43, 44

vitamin, 138, 140, 143

water soluble (-s), 186
white oak, 65
white pine, 63, 65, 67

yield table, 181-2